RECOMBINANT DNA
THE UNTOLD STORY

RECOMBINANT DNA

THE UNTOLD STORY

JOHN LEAR

CROWN PUBLISHERS, INC. / NEW YORK

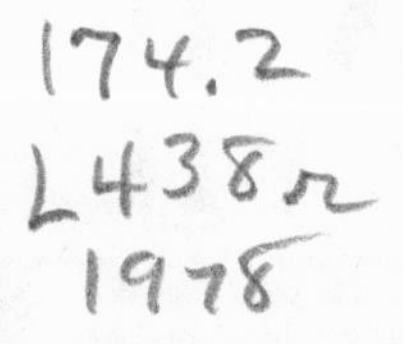

FOR MARIE

Inquiries should be addressed to Crown Publishers, Inc., One Park Avenue, New York, N.Y. 10016
Printed in the United States of America
Published simultaneously in Canada by
General Publishing Company Limited

BOOK DESIGN: SHARI DE MISKEY

Library of Congress Cataloging in Publication Data

Lear, John.
Recombinant DNA.

Includes index.

1. Recombinant DNA. 2. Genetic engineering–Social aspects. I. Title. [DNLM: 1. DNA, Recombinant, QU58 L438r]
QH442.L4   174'.2   77-29158
ISBN 0-517-53165-8

PREFACE

Almost a half century ago, scientists discovered that all the physical characteristics of our bodies are passed from one generation to the next through a particular kind of molecule—the DNA molecule. And just as the human DNA molecule determines that each of us will be a person, so the DNA molecule of a cat determines that the cat will breed only cats, the DNA molecule of a horse determines that the horse will breed only horses, the DNA of a housefly determines that a housefly will breed only houseflies, the DNA molecule of a maple tree determines that the maple tree's seeds will always grow into maple trees, the DNA of a microbe that causes pneumonia determines that the pneumonia microbe will breed microbes that cause pneumonia and not viruses that cause measles, and so on *ad infinitum.*

Nine years ago a graduate student at Stanford University proposed a simple method of taking the DNA molecule from one species of life and splicing it onto the DNA molecule of any other species of life. Other scientists since then have refined and changed his method in a variety of ways which today are known collectively as recombinant DNA technology.

Recombinant DNA technology is a unique tool for helping us do two kinds of things:

To explore the genes in the DNA that make us what we are, to discover how they work together to keep us alive and well (or, in some cases, sickly or crippled), and perhaps someday teach us how to treat or even prevent the diseases we inherit.

To splice particular genes into the DNA of selected bacteria, which will reproduce the DNA and manufacture very large amounts of such things as hormones, blood fractions, and antibodies that fight infections.

In doing such things this way, there may be some risk of changing the microbes from harmless creatures to dangerous ones, or of creating hybrid viruses that might start epidemics. The scientists who do recombinant DNA experiments have an obligation to tell us candidly what to expect and to protect us from harm.

Unfortunately, the scientists are split into two violently antagonistic schools of thought about the seriousness of the risks and the likely value of the benefits. The fury of the battle and the nature of some of the booby traps that have been and are still being used have frightened lawmakers away from their responsibility to protect us. The decision cannot be put off much longer without inviting disastrous consequences to science and to our democratic society.

This book is the untold story of how we got into this tragic situation and what alternatives we have to assure our common survival.

ACKNOWLEDGMENTS

The original impetus for this book came from the late Paul Nadan, one of the most gifted senior editors at Crown Publishers. To him these pages constituted only a small sector of a great battlefield where he fought for the freedom and dignity of everyman. He died with his boots on, ambushed on a lone patrol, just a few hours after clearing final galley revisions with me by telephone. In saluting him I am mindful of his ever-present courtesy; he would be unhappy were I to neglect thanking all those others at Crown who helped to shape the book into a message worth the attention of a democratic people.

On a quite different level of the spirit, I must also express my appreciation to the two Nobel laureates and several professors of microbiology who would not talk to me about recombinant DNA technology. If their behavior had not aroused my curiosity, I might not have dug as hard as I did to find out why, and the quality of this book would have been diminished accordingly.

I am reluctant to name the many scientists who were helpful and kind, lest they be held responsible for the candor with which personalities and events are described in these pages. But my private debt to them is large indeed.

Otto Rosahn and Sonya Staff will understand my gratitude to them. Likewise, Alfred and Phyllis Balk. And Esther Leeds.

How does one thank the Ford Foundation for a travel and study grant? If a respectful yet affectionate bow will do, I incline my head.

The data in the MIT Oral History Project files were open to me to a limited degree. The project staff was writing a book of its own, and that took priority. But I was able to confirm a number of findings independently arrived at.

Susan Schroeder's typing was as clear and clean as mine was ramshackle.

The unquenchable faith of my wife, Marie, kept me sane in a setting of madness whose intensity I would not believe had I not lived there for twenty months.

PROLOGUE

I

On the afternoon of Monday, June 28, 1971, Robert Pollack, a thirty-one-year-old microbiologist on the research staff of the Cold Spring Harbor Laboratory, Long Island, made a telephone call that would fundamentally change the relationship of American science to the democratic society that shelters it.

* * *

The change would frighten some people. Others would welcome it, knowing that when change stops life stops too. Even the orientation of earth in relation to distant neighbors of our homestead star, the sun, shifts imperceptibly from minute to minute as the planet goes spinning along, wobbling slightly on its axis as a top does, inscribing invisible circles in the sky.

One of those circles takes 26,000 years to complete. During that period the northern pole points to one star after another, and the place at which the sun rises over the eastern horizon on the first day of spring moves slowly along the plane of the ecliptic (in which the earth travels around the sun) from one constellation of the zodiac to the next. About 2,200 years pass between the time the equinoctial sunrise enters one constellation and the time it enters another.

Astronomers and mathematicians have known of this phenomenon for thousands of years. But only recently have a few science historians been gathering evidence that primitive peoples have known of it too and have perpetuated accounts of it in myths that recur in many languages—all sharing the central belief that with the passing of one star age a long established order of human affairs crumbles and is supplanted by a different way of life.

"A new great order of centuries is now being born," the Roman poet Virgil wrote at the start of the Christian era. On the day of the spring equinox the sun rose at the point where before dawn the constellation Pisces could be seen. Pisces, the fish. And the fish became and ever after remained a symbol of Christianity.

Before entering Pisces, the point of sunrise on the opening day of the growing season had moved through the stars of the constellation Aries. Aries, the ram. By biblical account, it was in that period when the Hebrew prophet Moses, using a ram's horn to call his people to prayer, had difficulty in weaning them away from the "Golden Calf," which was Taurus the bull, the constellation that Moses (an astronomer, according to Sigmund Freud) knew had just been abandoned by the sun.

At present, the equinoctial sunrise is moving toward the constellation Aquarius, an event that has already been anticipated in the musical comedy *Hair*. According to the star-age calendar, then, a revolutionary change is due in our lives.

Whatever connections there may be between heaven and earth to explain the coincidence, a change of great proportions is indeed engulfing us. After thousands of years of using sexual mating to manipulate the evolution of animals and plants that feed and otherwise serve us, science has learned how to create, in the laboratory, hybrid species of deoxyribonucleic acid (DNA) molecules—the ones that carry detailed instructions for the reproduction of whatever organism the molecule belongs to, be it a flower, a grass, a tree, an insect, an earthworm, a bird, a whale, a mouse, or a human.

The DNA molecules are, as everyone knows, the residences of the genes. In theory at least, genes from any organism whatever can now be spliced onto the genes of any other organism, no matter how unrelated those organisms may be in nature. As one of the pioneers of gene-splicing techniques has put it: "For the first time, there is available a method which allows us to cross very large evolutionary boundaries, and to move genes between organisms that have never before had genetic contact." It is obviously impossible to tell, at this point in time, where the new age (some have called it "The Age of Playing God") will take us.

II

"A method which allows us to cross very large evolutionary boundaries."

How large?

So large that they must be measured in millions or even billions of years.

Events that span such grand sweeps of time call for poetry, and it emerged in prose form in the *New England Journal of Medicine* in 1971, 1972, and 1973 under the signature of Dr. Lewis Thomas, a research physician, a member of the National Academy of Sciences, and, at the time his words appeared in a book titled *Lives of a Cell,* president of the Memorial Sloan-Kettering Cancer Center in New York City. Here are two singularly pertinent lines:

> The uniformity of the earth's life, more astonishing than its diversity, is accountable by the high probability that we derived, originally, from some single cell, fertilized in a bolt of lightning as the earth cooled. It is from the progeny of this parent cell that we take our looks; we still share genes around, and resemblance of the enzymes of grasses to those of whales is a family resemblance.

Even the approximate date of that postulated lightning bolt remains mysterious, but scientists expert on such matters place it anywhere between 4.6 billion years ago, when the earth was formed, and 3.5 billion years ago, the time of creation of rocks containing fossils of early descendants of the original parent of all of us and all of the plants and animals on which we feed.

The postulated great-great-great granddaddy's great-great-great granddaddy cell is now being called the "progenote." A team of genealogists of submicroscopic life (Professor Carl R. Woese, of the University of Illinois, and Professor George E. Fox, of the University of Houston, are its senior members) has published a rationale for the belief that from the progenote's immediate offspring arose three lines of progeny, two of which have remained single-celled creatures ever since while the third line found ways to slowly integrate themselves into multicelled beings.

Much the simpler in genetic structure and therefore presumably the older of the two single-celled branches of the great family of earthly life still live in airless places as diverse as the human gut and the hot springs of Yellowstone National Park, taking in carbon dioxide and hydrogen, giving off methane. Almost nothing is known about these methanogens, or, as they have also been called because of their apparent venerability, "archaebacteria."

But whole libraries have been compiled about the second one-celled line of earthly inhabitants, which include the blue green algae and all bacteria except the methanogens. That this line has been here for at least 3.5 billion years was established by Professor Elso Barghoorn of Harvard University, who used a diamond saw to slice samples of African rock of that age so thin that light could pass through them. He turned the lenses of a microscope on those fragile wafers and saw many kinds of primitive life well preserved. They were similar to the blue green algae that exist today in ancient stromatolites—uniquely layered rocks.

The layers of the stromatolites are actually bound together by the intertwining algae, which excrete oxygen and probably are responsible for enabling the evolution of oxygen-breathing animals including ourselves. For astronomers say the environment of earth at the time of its creation would not have allowed the presence of oxygen then or for a long while afterward (but would, of course, allow the archaebacteria to thrive).

The fossil record in the rocks says that the algae and the bacteria were the earth's only inhabitants during the planet's first 2.3 to 3 billion years. The earliest fossils of plants and animals are only 600 million years old.

DNA splicing experiments have already joined the inheritance-governing molecules of bacteria and lower animals such as fruit flies, silkworms, and toads. Billions of years of time have been crossed in a single stroke. The intent of the scientists who have done these experiments is to better our lives by discovering ways to intervene in treatment or possibly even cure and prevention of hereditary disease. But can we be sure that these evolutionary leaps will always carry us forward? May there not be some risk of falling backward?

The chairman of the board of editors of the *Proceedings of the National Academy of Sciences,* Professor Robert L. Sinsheimer, who recently became chancellor of the University of California at Santa Cruz after many years as head of the biology division of California Institute of Technology, has repeatedly expressed concern over the prospects. Too many episodes in the story of life on earth are not well enough understood, he says, to justify moving too fast or without sufficiently strict precautions.

Most of us who are not scientists are rather poorly equipped to follow the logic behind his reasoning. For we and our forebears for generations past have looked upon life as consisting of two categories—plants and animals. It was an entirely reasonable view until the 1930s. Up to then, virtually all research in genetics was done with plants and animals. One breeding experiment could include no more than a few thousand subjects, usually no more than a few hundred. A single life cycle lasted for weeks, months, or years. The accumulation of meaningful data was discouragingly slow and cumbersome.

But on the molecular level of life, plants and animals are more alike than they are different from each other. And it is on the molecular level that most genetics research of today is done with bacteria and viruses anywhere from ten to ten thousand times smaller than a single human cell. A generation spans only twenty minutes. By the end of a day, one parent organism can have millions of offspring.

Fortunately for us nonscientists who have poor memories, we need to identify only a few creatures in order to follow the mainstream of events. The most ubiquitous is *Escherichia coli,* a species of bacteria isolated from human feces in 1885 by a German pediatrician, Theodor Escherich, and named for him and the bacterium's normal habitat: the colon of warm-blooded animals, including man. Nicknamed *E. coli, Escherichia coli* is a rodlike organism two wavelengths of light long and one wavelength wide. Next to it in historical importance are the bacteriophage—specialized viruses—that live symbiotically on *E. coli.* When visualized by the microscope, phage look like tadpoles, attached by their tails to bacteria, which loom alongside like walruses.

These invisible organisms—many times more numerous than the peoples, animals, and plants of earth—live on one side of the most fundamental dividing line in life, the prokaryote side. Prokaryotes are one-celled organisms without a nuclear membrane but with a single very long and monotonous-looking string of DNA (the chromosome) floating freely within the cell along with one or more submicroscopic rings of autonomously replicating DNA, some packets of reserve resources, and the ribosomes, which are miniscule factories where amino acids are assembled into proteins according to DNA instructions carried and transferred by a less celebrated but equally vital molecule named ribonucleic acid (RNA).

On the opposite side of the great divider from the prokaryotes are the eukaryotes, including all the plants (even the fungi) and all the animals from birds to fish to insects to us. Eukaryotes (with the exception of the protozoa) are multicelled creatures, each cell having its complement of ribosomes in the cytoplasm but being organized around and controlled by a nucleus containing a set of complex strips of DNA (the number varying with species) and some other components, including the mitochondria, which have the most to say in deciding whether we are quick or dead.

This is how the physician-poet Lewis Thomas described the crucial role of the mitochondria in our lives and set forth what seems to be the likeliest supposition about the way they got into position to play it:

> At the interior of our cells, driving them, providing the oxidative energy that sends us out for the improvement of each shining day, are the mitochondria, and in a strict sense they are not ours. They turn out to be little separate creatures, the colonial posterity of migrant prokaryocytes, probably primitive bacteria that swam into ancestral precursors of our eukaryotic cells and stayed there. Ever since, they have maintained themselves and their ways, replicating in their own fashion,

> different from ours. They are as much symbionts as the rhizobial bacteria in the roots of beans. Without them, we would not move a muscle, drum a finger, think a thought.
>
> Mitochondria are stable and responsible lodgers, and I choose to trust them. . . . I like to think that they work in my interest, that each breath they draw for me, but perhaps it is they who walk through the local park in the early morning, sensing my senses, listening to my music, thinking my thoughts.

All about us biologists trained to do so recognize examples of the symbiotic relationships that prokaryotes have arranged on the road of evolution into eukaryotes. These are the modern progeny of intermediate experiments in survival, slow and patiently devised innovations that are being bypassed by the gene splicers, who can only guess at the reasons why particular ingenuities were necessary and desirable at particular navigation points on the ocean of time.

A familiar case involves the bacteria that live in our intestines and manufacture vitamins essential to our well-being. They are partners of ours, and must be present inside us in right numbers and proportions if our food is to nourish us adequately. A less widely known example is a lichen common in the northeastern United States. Called the British soldier lichen, it is a partnership between a fungus, which provides shelter, and algae that convert sunlight into food for the lichen. A more ingenious example of symbiosis is a mollusk named *Elysia,* a green sea snail that lives like a plant. *Elysia* draws carbon dioxide from the air and turns it into food with the help of sunlight. When it is young, the snail eats seaweed and sucks out the chloroplasts, the parts of seaweed cells that perform photosynthesis. The snail does not digest the chloroplasts but somehow moves them to places in its digestive system and uses them to perform photosynthesis for itself. The snail does not eat for the remainder of its life, but lies in the sun and enjoys itself. There is a widely shared belief among qualified scientists that all eukaryotes have a collective assortment of prokaryote ancestors. But the pattern of evolution is an immense jigsaw puzzle, and no one yet understands precisely how the pieces fit together.

Professor Sinsheimer believes there must be a very good reason why exchange of genetic information between prokaryotes and eukaryotes is not known to occur naturally. He therefore assumes the existence of an invisible barrier, which can be breached only at the breacher's peril—a wall of knowledge to surround the tree of knowledge of the Book of Genesis.

Sinsheimer argues that even if we knew much more than we do, and were able to trace coherently the steps in the evolutionary process, we still could not blindly rely on the past as a guide to our present safety. The environments in which the original progression occurred, he points out, were far different from the environment of today. Besides, the speed of evolutionary change is being deliberately accelerated by the gene splicing, and the environment will have less time to adjust to the consequences of hybrid experiments than it had on the first go-round.

Few other scientists are willing to debate Sinsheimer on his chosen philosophical ground. Most of them aren't deeply concerned about the long-range effects of their work. Despite the many instances in which it has failed to appear in the past, they persist in a naïve trust that a technological fix will appear if and when it is needed. As Sinsheimer cannot prove that the barrier he supposes to exist actually does exist, they prefer to ignore him, or ridicule him, or try to chip away his reputation through innuendo in private conversation.

I bear personal witness to this last with a sense of sadness and shame. Experience has taught societies much older than ours to cherish and protect their Jeremiahs. If we cannot follow those precedents, we can at least keep the record reasonably straight. Sinsheimer's period as chairman of the board of editors of the *Proceedings of the National Academy of Sciences* has not been marked by footless speculations or other displays of irresponsible behavior. Moreover, Sinsheimer has not urged a total ban on gene splicing. He has simply pleaded that we give ourselves time to appraise potential safeguards and meanwhile not fall prey to the illusion that in so doing we are transgressing some vague inalienable right of free scientific inquiry.

Sinsheimer became involved in nucleic acid research very early in his career. He believed then and believes now that "wonderful results are to be derived from genetic engineering," including some "that may literally be essential for the survival of our civilization." At the same time, he sees a "darker potential for biological and social chaos" if the gene-splicing research is allowed to proceed headlong.

To the innocent layman who pays the gene-splicers' bills through his taxes, Sinsheimer's attitude must seem so reasonable as to be beyond debate. But laymen are not familiar with the scientific mind, which Sinsheimer, having spent the whole of his adult life in science, knows to be dedicated to the conviction that whatever work a competent researcher chooses to do is important and will ultimately be beneficial to mankind. "A major wrench to the thinking of many if not most scientists" will have to occur, he says, before they begin to accept what is for them a novel and painful conception—"that science can make the world a more dangerous place, that scientific advance could destabilize society and could even imperil the human future."

III

The number of molecules that could be fitted within a hollow pinhead the size of a period on this page is many times the number of people who presently inhabit the earth. Consequently, much of the art of splicing DNA molecules is necessarily accomplished outside the range of the human eye, even when the eye has the help of the electron microscope, which magnifies objects under its lenses 100,000 times actual size. The "em," as scientists have affectionately nicknamed this fabulous instrument, can see DNA only as a slender thread. The "em" likewise sees RNA as a slender thread. The same slender thread. There are no marks or characteristics to visually discriminate between the two, whose functions differ so greatly as to demand unequivocal definition.

How, then, does a modern gene-splicing laboratory run? On clever deduction as much as anything. By patiently waiting and watching to see what happens when something is done differently from before. An ingredient added or subtracted. Heat or radiation turned up or down. The acid-alkali balance shifted. A segment of DNA chipped off. A sequence of genes turned backward.

The process is guided by as many empirically compiled cookbooks as Grandma ever kept as kitchen aids. A researcher who wants to know the recipe of an experiment looks up the notes of an earlier scientist who successfully performed the experiment. There is only one large aberration in the analogy: Grandma either guessed at temperatures and oven times or relied on primitive thermometers and clocks, whereas the instruments that abound in molecule-splicing laboratories indicate exactly how many times a centrifuge spins in a minute, how often a flask shaker shakes, precisely how hot are the ovens and how cold the refrigerators, how dilute are the enzymes, how fine the granulation of the grinders, how intense the ultrasonic bombardment or the beam of ultraviolet light, how much radioactivity is emanating from isotopically labeled genes, how one DNA fragment trapped in a block of sensitized gel relates in size to each of the other 199 fragments that may be trapped on that same block of gel, and what statistical patterns the computers are discerning in the whole concatenation of technical complexities.

A lengthy recitation of the details of all the processes involved would bewilder everyone but specialists. But anyone who can count up to three and recognize four of the letters of the Roman alphabet—A, C, G, and T—will be able to follow the logic that is basic to DNA splicing. All living organisms are built of proteins. All proteins are built of amino acids. All amino acids are built of one or another combination of three of the four nitrogenous bases—adenine (A), cytosine (C), guanine (G), and thymine (T)—that are strung together by phosphates and sugars to constitute the DNA molecule. The DNA molecule is a double-stranded ring. Within it, at every point where an A appears on one strand the other strand must have a T, and opposite every C must be a G. Otherwise, the molecule falls apart.

The first clue to this most fundamental of all arrangements of the structure of living creatures was found by Professor Erwin Chargaff, an Austrian-born biochemist on the faculty of Columbia University. The following pages will tell something of his findings and show how other scientists built on them to win Nobel Prizes. Here it is enough to note that his credentials are ample to justify his participation, at the uppermost levels, in discussion of the manipulation of the basic mechanics of heredity.

Like Sinsheimer, Chargaff has taken a sweeping evolutionary view of DNA splicing. In a letter to *Science,* journal of the American Association for the Advancement of Science, he asked essentially the same question that Sinsheimer asked: "Are we wise in getting ready to mix up what nature has kept apart, namely the genomes (genetic totalities) of eukaryotic and prokaryotic cells?"

"Bacteria and viruses have always formed a most effective biological underground," the letter to *Science* continued. "The guerilla warfare through which they act on higher forms of life is only imperfectly understood. By adding to this arsenal freakish forms of life . . ., we shall be throwing a veil of uncertainties over the life of coming generations."

Up to that point, the positions taken by Sinsheimer and Chargaff were identical. However, Chargaff is by nature as caustic as Sinsheimer is gentle, and the contrast asserts itself in the close of Chargaff's letter:

"Have we the right to counteract, irreversably, the evolutionary wisdom of millions of years, in order to satisfy the ambition and the curiosity of a few scientists?"

Chargaff's final sentence was a scorching prophecy:

"The future will curse us for it."

IV

From its very beginning, the science of genetics has been subject to fits of faulty communication, not only between practitioners of the science and the people at large but among the involved scientists themselves.

The first fit totally eclipsed the genius of the Augustinian monk Gregor Johann Mendel, who founded the science by crossbreeding long-stemmed and short-stemmed peas and then red-flowered and white-flowered peas in the garden of the monastery at Brunn, Austria, for eight years. Mendel kept meticulous notes, which he summarized in a report published in the *Proceedings of the Natural History Society of Brunn* in 1865. The report amounted to a statement of the natural laws governing the inheritance of characteristics from parents to offspring, and also set forth Mendel's assumption of the existence of "formative elements" capable of determining single heritable characteristics.

Mendel knew he had made an important discovery. He likewise knew that, being an amateur, he couldn't attract much attention on his own. So he sent a copy of his findings to a well-reputed Swiss botanist, Karl von Nageli. Von Nageli, having his own ideas about the nature of inheritance, brushed the obscure monk aside. Mendel died in 1884, completely neglected by the scientific world. The highest honor he acquired in his lifetime was appointment as abbe of the monastery.

Only because a conscientious German compiler listed Mendel's report in an extensive scientific bibliography was Mendel's seminal work saved for later discovery by other botanists and plant breeders who, through their own researches, reached the same conclusions that Mendel earlier came to. The long delayed knowledge finally passed from Europe to the United States, where, in the early 1900s, biologists Walter Sutton and Thomas H. Morgan theoretically located Mendel's invisible "formative elements" on long, rodlike bodies, which the microscope revealed within living cells. These rods were given the name, chromosomes. At times of cell division, they were seen to pair off and divide in much the same way that Mendel had imagined his "formative elements" would do. By bombarding fruit flies and corn with X-rays in independent experiments, Hermann J. Muller and L. J. Stadler next showed that the "elements" on the chromosomes could mutate individually and so account for varied traits within species as well as evolution of species. Later on, George Beadle and Edward Tatum used a common red mold that grows on bread to demonstrate that a single "element" (it meanwhile had come to be called a gene) does its work by controlling a single enzyme. Enzymes take their name from the Greek word for leaven because, among other things, enzymes account for the leavening of bread. They are naturally occurring organic substances that facilitate the myriad metabolisms involved in growth and other life processes of plants and animals including ourselves. According to classical genetic theory, each enzyme is responsible for the construction of a particular protein whenever the appropriate gene calls for it.

Three years after Mendel unwittingly buried his magnificent

insights in the *Proceedings of the Natural History Society of Brunn,* a different line of genetic communication was attempted and almost immediately broken. In Tübingen, Austria, a traveling Swiss scientist, Friedrich Miescher, isolated a bundle of Mendel's "formative elements" without knowing that Mendel had suggested their existence. Miescher was studying the nuclei of cells. Under the microscope, it became obvious that the nucleus of the pus cell made up a disproportionately large part of the cell. Miescher therefore collected pus cells from oozing and stinking bandages stripped from purulent wounds in the Tübingen hospital. After soaking the pus off the bandages, he dissolved it with pepsin, the digestive enzyme of the stomach. A white powder was left. Unlike other then known proteins, the powder was not soluble in water or in dilute acid. But it did dissolve in dilute alkali. A further sign of the stuff's unusual nature appeared when chemical analysis showed it to contain phosphorus, a source of the energy that keeps life going. Miescher named his discovery "nuclein," and described it scientifically in a paper he sent to Professor Felix Hoppe-Seyler, the publisher of *Hoppe-Seyler's Journal of Medicinal Chemistry.* Hoppe-Seyler held up publication of Miescher's information until Hoppe-Seyler could confirm the data himself and simultaneously publish a paper of his own as well as companion papers by two of his students, who determined that "nuclein" was present not only in pus cells but in the red cells of animal blood, in yeast cells, and in casein from milk. How much greater benefit might have accrued from participation of a larger number of scientists in the confirmatory experiments can only be guessed.

However events might have rearranged themselves in a more open context, there actually occurred a lapse of almost sixty years before "nuclein"—which in the meanwhile acquired a new name, nucleic acid—became a major factor in genetic advance. And the new impulse originated not as an imaginative leap in an individual mind but as a reaction of the medical profession in Europe and America to the appalling death toll of pneumonia at the turn of the twentieth century. In the United States pneumonia then ran a very close second to tuberculosis as a destroyer.

Professor Arnold W. Ravin, of the University of Chicago, has written engagingly of this period when genetic research merged with larger concerns of society. Part of the historical setting was the then recent discovery that microorganisms were the causative agents of infectious diseases. A new branch of science, medical microbiology, had emerged with the objective of identifying and controlling these killers. By the 1920s, several significant pieces of knowledge had been put together. Of the numerous forms of pneumonia, the most common cause of death was the pneumococcus, a spherically shaped bacterium that infected the lungs. Virulent pneumococci, those that produced fatalities when injected into susceptible animals, secreted around themselves a slimy coat or capsule. Different types of this microbe could be identified by the specific chemical content of the coat. Typing was an important aspect of clinical treatment because antibodies attacked only a specific coat and no other. In those pre-antibiotic days, the control of pneumonia therefore depended on awareness of the kinds of coats that covered the prevalent microbes. The number of distinct coats turned out to be very large. Not only was every coat chemically unique, but the organism within the coat nearly always transmitted to its descendants the ability to manufacture that coat alone. As sometimes happens among laboratory-bred organisms, however, the progeny of originally coated pneumococci occasionally lost the ability to make a coat. When those uncoated individuals were injected into experimental animals, no infection resulted.

At that point Dr. Fred Griffith, a medical officer in the British Ministry of Health, made history. Studying the difference in virulence of various types of pneumococci, he isolated colonies of bacteria from the sputum of patients with lobar pneumonia, grew pure cultures of those organisms, and injected the bacteria into mice. Those pneumococci that had rough coats (R pneumococci, he called them) caused no disease in the mice, but smooth-coated pneumococci, (S pneumococci) killed the mice. To learn whether harmless bacteria could revert to virulence, he killed some S pneumococci with steam and injected them along with live R pneumococci into other mice. Many of those mice died. From the lungs of the dead, Griffith recovered many living S bacteria. Within the experimental animals, the harmless R bacteria had somehow been transformed into virulent S bacteria.

What Griffith had observed was a natural recombination of DNA. He couldn't possibly have realized that because the vital function of DNA had not yet been identified. Nevertheless, he described his experiments so faithfully that other scientists who followed him could dig out further information and with it determine the truth about the way in which hereditary characteristics are passed from one generation to the next.

"So surprising were Griffith's findings," Professor Ravin says, that "few microbiologists trusted them"—least of all Oswald Avery, a physician and microbiologist at the Rockefeller Institute for Medical Research in New York City (now Rockefeller University), where Avery had been carrying on his own studies of pneumococci. In Avery's laboratory, however, was a young Canadian, Martin Dawson, who had confidence in Griffith's work. Largely on his own, Dawson not only confirmed the phenomenon reported by Griffith but showed that the mice were not required for its replication. Griffith's results could be reproduced just as well by mixing R and S pneumococci in a sterile test tube containing nothing more than a nutritional medium on which the bacteria could grow. Lionel Alloway, another young collaborator of Avery's, took up the quest then and found that the heat-killed S bacteria could be replaced by a crude, cell-free extract derived from those bacteria and still duplicate what Griffith had seen in the test tube.

Professor René Dubos, who worked with Avery at the Rockefeller Institute, recently published a book about the experience. Titled *The Professor, the Institute and DNA,* it confirms Ravin's account. Avery's original skepticism finally gave way, and for more than a decade (1932–1944) he searched for the constituent of S bacteria that was responsible for transforming R bacteria. This constituent, he realized, was capable of directing hereditary change. It carried genetic information. It was like genes. And in 1944 Avery and two senior associates at Rockefeller (Colin Mac Mcleod and Maclyn McCarthy) published a report stating that the DNA and it alone was the material with the genetic properties.

If Avery and his research team had won a Nobel Prize for their work, as they deserved to, a serious misunderstanding about present-day gene splicing might have been avoided. Nobel Prizes are at least as effective as schoolbooks in spreading knowledge of scientific discoveries. Had a prize been given to the Rockefeller Institute group, everyone soon would have become familiar with the fact that nature has been making hybrid DNA molecules for years. Some of them may be hazardous to us humans, but most of them are almost certainly not. The danger doesn't lie in the recombination per se, but in the choice of the molecules that are to be spliced together. Hybrids that aren't natural are the scary ones. They ought not to be called simply "recombinant DNA molecules" but "novel recombinant DNA molecules."

Seen in hindsight, the omission of a Nobel Prize for Avery and his research team seems incomprehensible. But in spite of the decisive quality of the experiments, there was so much unwillingness among scientists to accept the startling DNA concept that the Swedish Academy judges probably felt that a prize could be interpreted as an attempt to quash dissent. Not until eight years later were doubters finally silenced by a piece of evidence from the Carnegie Institution of Washington laboratory at Cold Spring Harbor, Long Island. There Alfred D. Hershey and Martha Chase put separate radioactive labels on the protein coat of a virus and on the DNA inside the coat before attempting to infect a bacterium with the virus. Only the

DNA entered the bacterium, and the bacterium was infected.

During the eight years of bickering, there were, of course, many scientists who accepted the Avery team's conclusions. They concentrated on learning what DNA consisted of, and how it was put together. Chargaff and his students at Columbia hunted hidden relationships among the fragments of DNA they analyzed. They found two such, both involving the nitrogenous bases: adenine (A), cytosine (C), guanine (G), and thymine (T). The number of As was always close to the number of Ts, and the number of Cs was always close to the number of Gs. Chargaff and his students also noticed that the proportion between As + Ts and Cs + Gs varied with the source of the DNA. Some species had more As + Ts than Cs + Gs while others had more Cs + Gs than As + Ts. But the As = Ts and Cs = Gs equation remained undisturbed.

Sinsheimer was just starting out on the road to becoming a scientist when news of the findings at Columbia began to circulate in the scientific community. He had completed his studies at Massachusetts Institute of Technology and had moved to Iowa State University, where he chose for his analysis a very small harmless virus, ØX174, which attacks *E. coli.* ØX174 is still being studied in the United States and Europe. Its DNA was recently shown to consist of more than 5,000 different segments: more than 5,000 As, Cs, Gs, and Ts strung out in various combinations. But in the late 1940s and early 1950s, Sinsheimer didn't know whether his test-tube results would be valid in real life. To find out, he took a sabbatical with one of the greatest figures of any period of genetics—Professor Max Delbruck of the California Institute of Technology. Delbruck had taught himself the niceties of infecting bacteria with viruses. Only if the DNA was virile would infection occur. Chargaff's formula—As = Ts, Cs = Gs—was a natural subject for discussion between the two men. They agreed that the formula would not allow the DNA molecule to have three strands. Nobel laureate Linus Pauling at that time was trying to perfect a three-stranded model of the molecule. Together Sinsheimer and Delbruck visited Pauling and pleaded for a shift in his stance. They failed in their mission. Had they succeeded, Pauling today might hold three Nobel Prizes instead of two and Dr. James D. Watson might never have had occasion to write one of the most bitterly debated books in the history of science, *The Double Helix.*

Watson was able to write *The Double Helix* because he and his British collaborator in DNA model building, Francis Crick, took Chargaff's formula seriously. The model Watson and Crick fashioned took the shape of a double spiral stairway with two railings made of sugars and phosphates connected by regularly spaced steps that split in half, one-half being attached to each railing. The halves were the

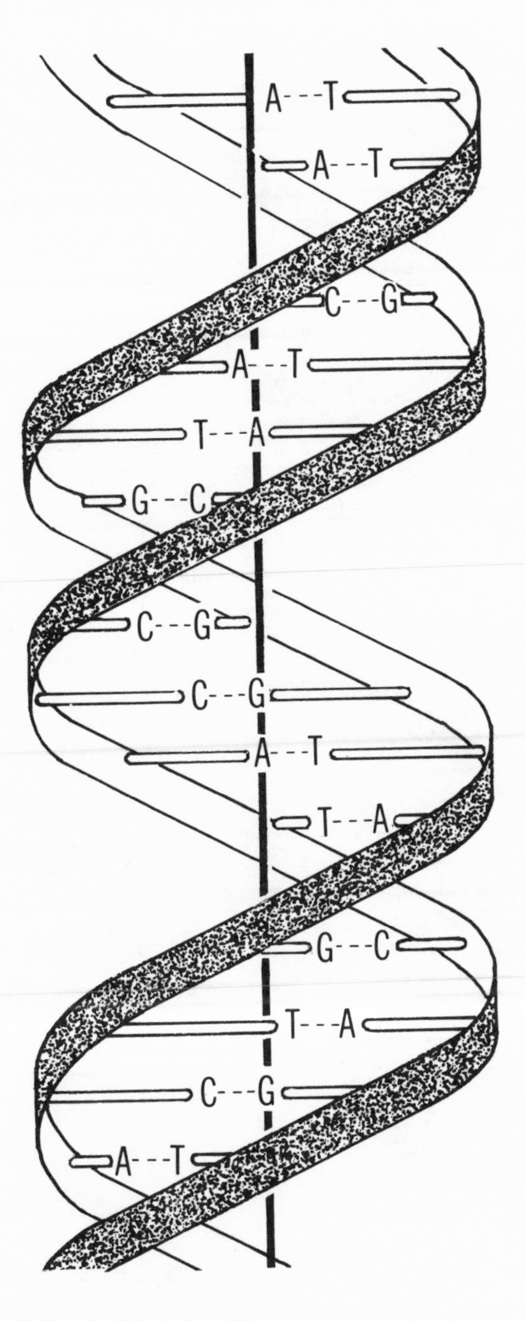

A schematic illustration of the double helix. The two sugar-phosphate backbones twist about on the outside with the flat hydrogen-bonded base pairs forming the core. Seen this way, the structure resembles a spiral staircase with the base pairs forming the steps.

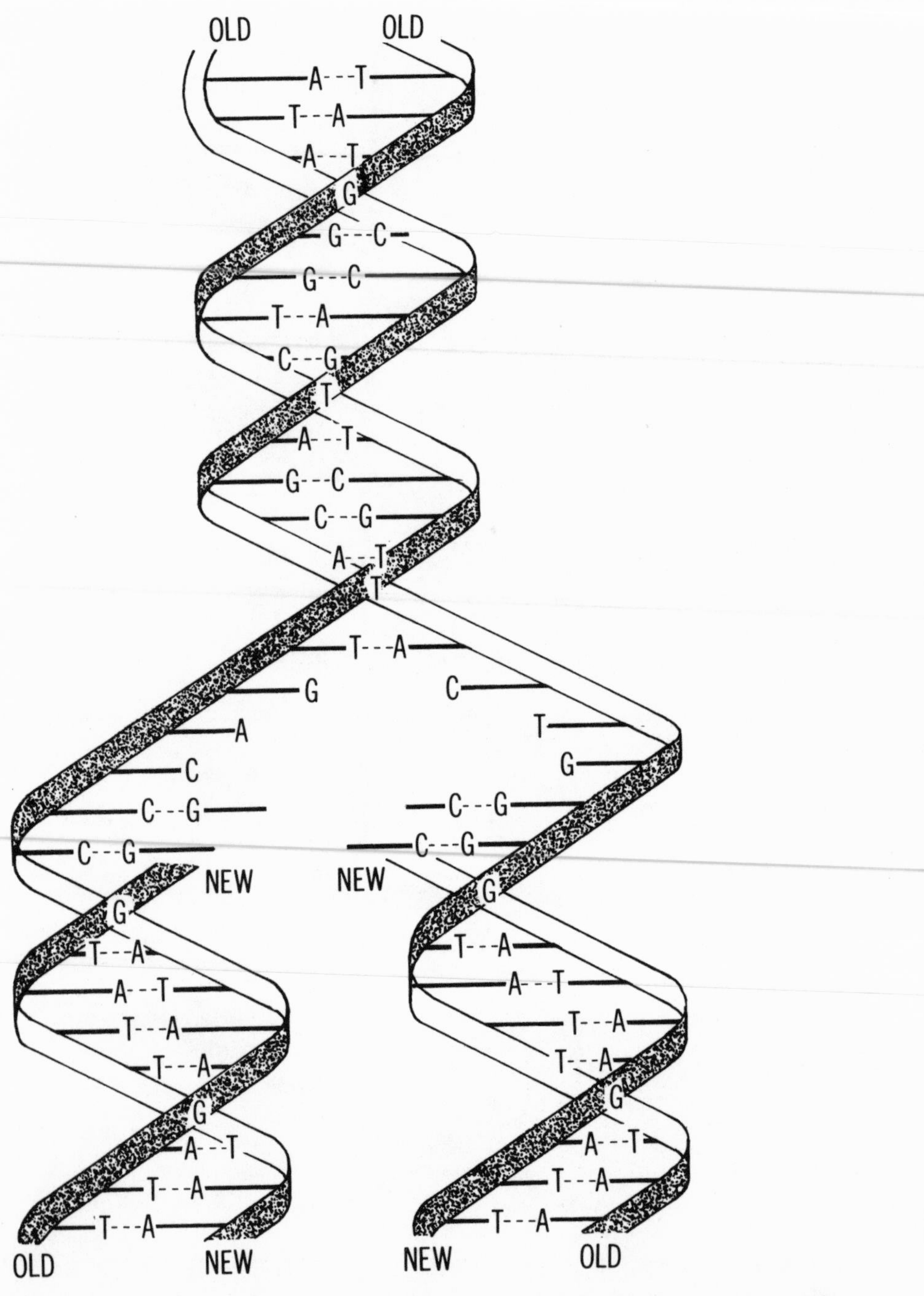

The manner envisaged for DNA replication, given the complementary nature of the base sequences in the two chains.

nitrogenous base pairs. As long as one-half step was an A and the matching half of that step was a T, or one-half step was a C and the matching half was a G, the stairway held together.

The model was especially elegant because it provided a simple explanation for what happens when the molecule replicates. The stair-steps divided and each railing with its half-steps became a template on which a complementary railing and half-steps would form.

The model was only a model, however. How closely it resembled the living DNA molecule no one could tell. And four more years went by before Professor Arthur Kornberg, of Washington University at Saint Louis, demonstrated in the test tube that a previously unrecognized enzyme, DNA polymerase I, performs a copying job that allows DNA to unzip down the middle, separating the stairway railings in readiness for replications. In the very next year after Kornberg's Nobel Prize-winning discovery of DNA polymerase I, two Harvard University researchers, Matthew Meselson and F. A. Stahl, showed that what happened in Kornberg's test tube also happened in living bacteria. No reasonable person could any longer doubt that DNA directs the inheritance of the characteristics of life. How the directions are carried out remained mysterious.

Almost exactly a decade before Kornberg's discovery, Joshua Lederberg, a precociously bright first year medical student at Columbia University, got bored with the repetitive details that prospective physicians must store in their heads and began to dabble in the genetics of the red bread mold that Beadle and Tatum had used in shaking genetic science out of the doldrums that followed the long spurt of learning from the genes of fruit flies. Quickly the dabbling took a serious turn, and Lederberg quit medicine to seek a doctorate in science with Tatum at Yale.

Tatum had become curious about the stories bacteriologists were telling of occasionally seeing under their microscopes pairs of bacteria suggestively close to each other. Were those bacteria mating? Everyone thought not because bacteria had been examined intensively for signs of sexual organs, and none had ever been found. Nevertheless, Tatum couldn't be sure, and the question was worth a definitive answer. Whatever had been observed obviously didn't happen very often, but if it should turn out to be sexual activity the possibilities for genetic research with bacteria would be enormous because of the biological simplicity of the organisms in comparison even with fruit flies, let alone animals, and also because of the extremely rapid rate at which bacteria reproduce.

Tatum asked Lederberg to investigate the situation and gave him two mutants of the K12 strain of *E. coli* to culture and observe. K12 had been isolated in 1922 from the feces of a woman patient who was being treated for diphtheria in the Stanford University hospital. Progeny of the original isolate had been bred through a quarter of a century in laboratories at Stanford, where it was used to teach would-be physicians how to diagnose diseases by identifying the causative agents. Lederberg brought worldwide fame to K12 by discovering, through a stroke of seldom equaled luck, that it was not only very sexy but was one of the few strains of *E. coli* that had sexual congress.

Chance intervened still further in the development of bacterial genetics by placing a laboratory refrigerator containing a vial of K12 within convenient reach of the workbench of Lederberg's first wife, Esther, at a strategic moment in her career as a scientist. On progeny of the original Stanford isolate that she happened to pick off the refrigerator shelf she found bacteriophage that no one had seen before. She called it the lambda phage (lambda is the Greek letter L), having no suspicion of the role it would later play in the construction of hybrid DNA molecules not known to exist in nature.

Out of the work of the Lederbergs and many others came a school of research whose influence was so dominant over so long a time that three of its founders shared a Nobel Prize in 1965. They were Drs. Delbruck, Hershey, and Salvador Luria. The school that brought them honor was called the "phage school," phage (from the Greek word for "eat,") being short for bacteriophage. As we saw earlier, bacteriophage are viruses. There are two classes of them: virulent and temperate. Virulent phage attack and kill bacteria in minutes by taking over the bacterial reproductive apparatus for the purpose of reproducing viruses. Temperate phage have an option: they can kill instantly or they can enter a bacterium peaceably, sometimes carrying strange genes with them, and incorporate their own and the foreign DNA into the bacterial chromosome. In this latter position, the phage reproduce as though they were genes of the bacterium. To distinguish this process from the unassisted transformation of bacteria through transfer of raw DNA such as Griffith witnessed in 1928, the intervention of phage in the transfer of genetic information from one bacterium to another is called transduction and the phage is known as a transducing phage. At some indeterminate time after the phage becomes part of the bacterial chromosome, a descendant of the original invader may pull up stakes, kill the progeny of the original bacterium it is inhabiting, and on rare occasions carry off with it nearby genes of the bacterium.

During the summer prior to the bestowing of the Nobel distinction on Delbruck, Hershey, and Luria, the phages they used to advance genetic knowledge were put to work in a spectacularly simple way—to isolate a single natural gene for the first time. The work was done at

Harvard University and announced a month after the Nobel award, simultaneously with publication of a formal report on the experiment in the eminent British scientific journal, *Nature.*

The Harvard scientists on that unprecedented occasion used two different phages that attach themselves to the *E. coli* bacterium. Both take *E. coli's* lac operator gene—the one that controls fermentation of sugar in the bacterium—away with them when their symbiotic relationship with the *E. coli* ends. It is not a clean break involving only the lac operator gene; other genes are hauled along too.

The excess baggage was eliminated by using a fundamental distinction that exists between the lac operon DNA carried off by one of the lambdas and the lac operon DNA carried off by the other lambda. The direction in which the genetic sequence—the arrangements of As, Cs, Gs, and Ts—was read by one phage ran opposite to the direction of which the sequence was perceived by the other phage. The DNA was, of course, double-stranded, and wound in the usual way, like a spiral stairway. Unwinding and separating the strands was easily done with appropriate chemicals. Once the strands were parted, it was a simple matter to put one strand from each phage into a test tube together. Only the genetic sequences in the lac operator gene were mirror images of each other and therefore complementary. They came together and formed a new double helix. The remaining genes in the two DNA strands hung in loose ends in the test tube and were dissolved by an enzyme that affects only single-stranded DNA. The pure lac operator gene alone remained.

By all normal criteria of science, the Harvard achievement should have been celebrated jubilantly. But when the team of six experimenters, headed by Dr. Jonathan Beckwith, then thirty-three, announced its feat at a press conference, the mood was definitely downbeat.

"We do not have the right to pat ourselves on the back," one of the six declared. In the long run, he said, the successful experiment might "loose more evil than good on mankind." Beckwith added: "The more we think about it, the more we realize that it could be used to purify genes in higher organisms. The steps do not exist now, but it is not inconceivable that within not too long it could be used, and it becomes more and more frightening—especially when we see work in biology used by our government in Vietnam and in devising chemical and biological weapons." Another team member said, "The work we have done may have bad consequences over which we have no control." Drawing an analogy to the unhappy history of atomic energy research, he concluded: "The use by the government is what frightens us." Still another of the group foresaw possible perversions to evil political purposes, such as spreading genetic deformities among people perceived rightly or wrongly as enemies, or "phasing everybody into the same skin color, height, personality, making it all appear as an aid to humanity."

Robert Reinhold, who reported the interview from Boston for the *New York Times,* gave his opinion of the event in passing. The ambivalence of the experimenters as they stood tall on the margins of the future, he said, "typifies the growing unease among scientists over the social and political consequences of their work." He was grossly understating. More and more perversions of science were exposed during the next half decade.

It was during this galloping epidemic of fear and despair that the ability to splice DNA molecules at will emerged. An environment less friendly to any revolutionary change would be hard to imagine. It is likewise difficult to conceive a less scientific response to a situation than the one that came from the scientific community. Instead of quietly gathering facts and cross-checking for accuracy, the scientists broadcast unsupported claims and screamed imprecations at each other. While playing amateur high pressure politics as vigorously as they knew how, they derided the legitimate lawmaking efforts of professional politicians chosen for the job by a democratic people. Instead of acknowledging the obvious fact that society itself is an evolutionary organism that cannot hope to survive absence of adequate control over any part of its body, some scientists have made absurd charges of dictatorship and threatened to quit research if they were not left free to fix their own definitions of their responsibility. All this has put science and society out of joint. Sooner or later, reasonable perspective must be restored. In anticipation of that eventuality, there should be set down a record of what really happened as opposed to the myths that are now adrift. As the following pages will show, even the time and place of the beginning of the gene-splicing age have been obscured. Common belief puts the place in New England and the date in June 1973. The true time was some years before then, and the true site was California.

1 POLLACK HAD a chinful of bristling whiskers but no academic tenure, so his future career was vulnerable to ruinous retaliation should the phone call fail of its purpose, which was to prevent what Pollack thought was a dangerous biological experiment then being prepared 3,000 miles away by a celebrated scientist who was not only a full professor on the faculty of one of the country's most prestigious universities but a repeatedly honored member of the National Academy of Sciences as well: Dr. Paul Berg, professor of biochemistry and sometimes chairman of the biochemistry department at Stanford University Medical School in California. The purpose of the experiment was to create, at the molecular level of life, the first hybrid organism not known to exist in nature.

June 28, 1971, was the Monday of the third and final week of a Cold Spring Harbor tumor virus workshop in which Pollack was an instructor. He was a relatively new and decidedly junior associate of Dr. James D. Watson, the "Honest Jim" who had shocked the scientific world by publishing a racy account of his winning of a Nobel Prize with Francis Crick for putting together an accurate model of the DNA molecule.

While yet a graduate student, Watson had embraced the then unproved belief of his Ph.D. supervisor, Professor Salvador Luria, that viruses were not, as then commonly supposed, exclusive agents of disease and death, submicroscopic creatures that flitted about helter-skelter like mosquitoes and, after puncturing the walls of the cells of whatever bacteria, growing plants, or animals they happened to light on, went into action like automatic syringes, squirting their venom inside the cells with lethal effect. No, Luria had said, viruses were more often than not (although not always) friendly symbionts of man, protein-coated capsules of naked DNA, which they had the capability of transferring to the chromosomes of their targets, thereby influencing the course of evolution of all life on the planet. The viruses did not act willy-nilly but used great discrimination in their choice of targets, generally limiting themselves to one species while eschewing all others. A bacterial virus, for example, rarely sought out man or the other animals. Those viruses that produced tumors in animals did so because their DNA, on being added to the DNA of the animal cells, confused the hereditary signals and stimulated cells to proliferate when they should have remained idle.

Such was the background from which Watson had operated throughout his adult life. In describing his personal path to the Nobel Prize in his book *The Double Helix,* he had explained that Luria's will to understand the chemistry of viruses was what propelled Watson to Europe on the study tour that ended in his prize-winning collaboration with Crick in England. On Watson's assumption of the directorship of the famous old Cold Spring Harbor lab in 1968, it could have been taken for granted that Watson would institute an annual summer tumor virus workshop as a prominent feature of his regime. For by that time the opinions of his teachers on the nature of viruses had been amply verified, and one very small virus discovered in 1960 had become the darling of genetics research on a global scale. It was shaped like a doughnut, and possessed only five genes, one of which had been identified as a tumor control gene. Molecular biologists and biochemists everywhere were scrambling to discover how that gene did its job. The discoverer could very well become a Nobel laureate. The name of this virus was SV (for simian virus) 40, the fortieth virus ever to be found to infect the monkey family.

The Cold Spring Harbor Laboratory had been deeply involved in the SV40 excitement. Indeed, it was during the year when Watson took charge of the place that one of his senior research associates, Dr. Joseph Sambrook, demonstrated that the DNA of SV40 could enter, remain stable, and replicate within the cells of several different species of animals. That fact had come to the attention of Professor Berg, who determined to use it in his laboratory. He disclosed his intentions to only a few members of his research staff. The youngest and most inexperienced of these was Janet Mertz, who had whizzed through a four-year MIT course in three years and was still in gear when she hit California. In the early summer of 1971, Berg sent her to Cold Spring Harbor to attend that year's tumor virus workshop.

Dr. Robert Pollack was teaching a workshop class in animal cell culture, which is to say the proper care, feeding, and growing of cells prior to their infection by viruses grown for that purpose in the lab. During the first two weeks of the class, Mertz repeatedly interrupted his lectures to point out instances where Berg was using different methods in an unprecedented piece of work she was playing a part in. As the excited young woman confided more and more of the planning for the Stanford experiment, Pollack could only admire her professor's objective and the elegant biochemical procedures he had specified for the execution of his scheme. Berg's approach was a radically new approach to an old dream.

The dream begins with the fact that every human individual is a complicated bundle of 50,000 different traits, ranging from hair and eye color to robustness of constitution, body build, degree of flatness of the feet, and the ability to carry a tune. According to genetic theory, each trait is controlled by two genes, called alleles. That means there must be at least 100,000 genes in each of the billions of cells in our bodies. If scientists are patient enough, and ingenious enough in mimicking what they see nature do, they believe they will in time learn where all the human genes are located in relation to each other and how they collectively guide the myriad multiplications of a single cell until it becomes an intricately constructed body with appropriately proportioned arms and legs and fingers and toes and a head designed to integrate their activity in moving through a lifetime of vicissitudes until death occurs.

Long before the whole pattern of the human genetic network is discernible, pursuers of the dream expect it will be possible to move genes in and out—as we now plug and unplug telephones and electrical appliances—in order to correct mistakes that were made in transmission of the original instructions coded in the DNA molecule. For each mistake, there is now either a spontaneous abortion, a birth defect, a tendency toward a particular infirmity, or susceptibility to a particular disease.

It was because of the complexity of the multicellular human genome (the totality of the genetic information and control system) that the "phage school" of genetics had attracted so many scientists. And it was the ability of phage to sometimes transfer DNA from one bacterium to another that Berg hoped to mimic to get into the animal cells. But he knew the bacteriophages were too discriminating to be expected to undertake the task. So in 1965 he had shifted his field of study from bacteria to animal cells and hit upon the animal virus SV40 as his transfer agent.

Now Robert Pollack, before joining Watson's staff at Cold Spring Harbor, had been an assistant professor at the New York University School of Medicine. He had studied SV40 there for several years. He knew that SV40 not only caused tumors in several small animals but (although never identified as a cause of cancer in any human individual) caused cultures of normal human cells in the laboratory to take on the characteristics of tumor cells. Furthermore, SV40 produced those effects in eight out of every ten opportunities, compared, for example, to another very small virus, polyoma, which causes tumors in approximately one of every ten chances.

Before Mertz began her illuminating interruptions of his lectures, Pollack was already nervous about the activities at Cold Spring Harbor. In another part of the lab from where he worked, preparations were under way to replicate the DNA of SV40 on a large scale. Just having that much raw DNA around worried him. Mertz's remarks served as amplifiers of his worry.

Pollack was particularly alarmed by Mertz's account of Berg's proposed experiment because she said the professor intended to graft the DNA of SV40 to the DNA of the well-known virus, the lambda bacteriophage, to allow the lambda to carry the SV40 hybrid into *E. coli.* If everything worked according to Berg's plan, the SV40 hybrid, once inside the *E. coli* would multiply as the *E. coli* multiplied, once every twenty minutes. Many copies of the hybrid would thus be made available quickly for the SV40 in turn to carry into the lab cultures of animal cells.

Pollack couldn't believe that any responsible scientist who understood the ecology of *E. coli* would seriously contemplate putting the DNA of SV40 into *E. coli.* According to his calculations, a single pellet of naked DNA would be 100,000 times as infective as the dilute forms of SV40 normally encountered in laboratory situations. And if the SV40 DNA should escape confinement and somehow become free to invade the human body, it would enter by an unnatural route that would completely evade the body's formidable immunological barriers. Should Pollack remain silent on these points, it would be difficult if not impossible for him to continue to face his class, for he had used the potency of SV40 as a standard reference in exhorting the students to exercise great caution in their work.

Did Mertz understand the chain of life involved in the Stanford project? She insisted she did. But she didn't seem able to translate that chain of reasoning to fit Pollack's interpretation of the hazards.

Pollack's frustration grew with each passing day. Where did his personal responsibility lie? He felt strongly that academic protocol would require him to discuss the matter with Watson and let Watson decide whether a confrontation with Berg was justified. Watson, however, compensated for his intellectual brilliance by behaving toward his juniors like a benevolent despot; his subordinates responded, behind his back, by calling their common workplace the People's Republic of Cold Spring Harbor. Although he was unfailingly meticulous in maintaining the safety of his own laboratory, he might, if approached when his mind was galloping in another direction, say that the safety of Berg's lab was exclusively Berg's affair.

Had that been Watson's opinion, it would not have been Pollack's opinion. Pollack was constitutionally unable to disconnect his own moral fibers from the well-being of other individuals no matter how far removed geographically. The protection of any human anywhere from preventable pain or suffering was of direct ethical concern to him. I tapped the source of this uncommon attitude at our first meeting. In trying to relate his work to that of other biologists I knew, I had asked him how old he was. "I was born one year to the day after Hitler's stormtroops invaded Poland," he said. Pursuing the curious emphasis of that reply, I asked him to recall the most vivid memory of his childhood. Again he surprised me, saying, "I shall never be able to forget the uncontrollable weeping of both my parents when they heard that the families they had left behind in Europe were wiped out by the Nazis." Although he had been only five years old when World War II ended in the Pacific, only three when the fighting ended in Europe, he said he still could not listen for long to anyone speaking German. Mertz's story might have confronted anyone else with three alternatives. First, ignore it. Second, risk raising a stink with Watson. Third, call Berg. Pollack had no choice. He had to call Berg.

Before doing so, Pollack sought the advice of his wife, Amy. Amy is an artist, and sees the truth through uncompromising eyes. She urged her husband to do what he thought was right, regardless of consequences. To someone who later asked if she had any doubt about her position, she quoted from an obviously unscientific piece of literature: "Let them pee on you if they must, but don't let them tell you it is raining." And on Monday afternoon, June 28, 1971, the Monday of the third and final week of the cell culture class, Pollack telephoned Berg.

Although he had grown up within a few blocks of the house where Paul Berg had grown up twelve years earlier in the Seagate section of Brooklyn, Pollack had never met Berg either in boyhood or in manhood. The situation hardly lent itself to a familiar approach. But the emotional pressure bearing down on him left Pollack out of phase with the niceties of etiquette.

"Hello, Paul," he blurted when he heard the professor's voice on the other end of the line. "You've got a zealous follower here."

As it turned out, no explanatory introduction was necessary. Mertz herself had phoned her professor earlier and told him of her classmates' criticism of the SV40 hybrid experiment. He had told her he thought her classmates were crazy. But she had insisted on detailing the criticisms. After hearing her through, he said, "Well, when you get back here we can discuss it. I personally don't think there's very much if anything to worry about." When Pollack called later on, Berg expected a repetition of the dialogue with Mertz. What he got was a quite different, bluntly worded challenge.

"Do you really mean to put SV40 into *E. coli?*" Pollack asked. "Can't you put it in *B. subtilis* [a bacterium that inhabits the soil, and lives by digesting dead and decaying organic matter] or something else that dies in the gut?"

"Forget about it," the professor said. "We're taking care of it."

"How are you taking care of it? What controls have you got?"

"Look, Bob," Berg said. "I can design an *E. coli* that can't grow except when I push the button. I can make it streptomycin dependent. I can make it temperature sensitive. I can put in a whole series of mutations that render it incapable of surviving outside the lab."

"Maybe," said Pollack. "But the hybrid molecule may get out before your specially engineered *E. coli* dies. The minute it is out it is into another strain of *E. coli* that isn't rigged with all those safety features. It just won't do."

"That's silly," the professor said. "Just forget about it."

"It's dangerous," Pollack insisted. "Very dangerous. I don't think you should do it. The issue isn't just *E. coli.* This is an intrinsically dangerous line of work. We're all over our heads. You're assuming that infectivity is proportional to the things you're used to working with. But you're actually working with molecules whose infectivity is unknown."

Finally Berg said, "We aren't ready to do that experiment yet. Let me think about your objections."

Still not willing to abandon the field without a decision, Pollack invited the professor to deliver a special lecture to close the tumor virus workshop on the next Friday, July 2. Berg begged off. The preparation time would be too short. Pollack gave up then, and the conversation ended.

If Berg wouldn't deliver the closing lecture, Pollack had to. His working title was a provocative question: "Our research in introducing foreign DNA—is it anybody else's business besides the scientists who are doing the work?" His text presented two different points of view. One view: It is the scientist's business to protect himself and laboratory colleagues; experiments then can be as bizarre as any party to the proceedings can imagine. The other view: It is the scientist's business to protect society; then certain things are precluded no matter how safe the scientists may feel.

To which view did Pollack himself subscribe? When his students left Cold Spring Harbor for the 1971 Independence Day celebrations in their respective hometowns, they carried with them copies of a "set of comments and suggestions" that Pollack and Dr. Joseph Sambrook had drafted and signed "in response to a general but vague concern about the dangers of the work we do." This seven-page document, dated on the Monday of Pollack's phone call to Berg, had four parts. The first three were headed General Laboratory Practice, Biological Material within the Lab, and Laboratory Design. There was nothing very new or exciting about any of them. But the fourth part brought readers up short with a question that is seldom asked of science students: "Are there any good experiments using human cells and viruses that should not be done?" The answer the Pollack-Sambrook memo gave was:

> Even if you chose to, you could not experiment on human beings. You all accept, therefore, that a boundary exists past which you must not let your curiosity carry you. Work with somatic human cells in culture has proceeded without any apparent hesitation, so such work defines the inner or "doable" side of this boundary. Now a class of experiments with human germ cells (e.g., *in vitro* fertilization, cloning, introduction of "absent" genes, or gene therapy) is becoming possible. Will you place those experiments on the same side as experiments on skin-biopsy cells, or will you group them with experiments on people? You have mobilization time to consider this question *before* you begin the work.
>
> A second related class of experiments involves the reverse process: putting human genes or the nucleic acid of human viruses into cells of other species, or into prokaryotic cells. The ethical problem here is minor, but the dangers (e.g., of creating a human tumor virus that can grow inside a bacteria like *E. coli* which normally sits in the human gut) are immense.
>
> If you are going to work with either of these two classes of experiments, we suggest you surrender a portion of the scientist's right to follow his nose without regard to consequences. You ought to ask yourself if the experimental results are worth the calculable dangers. You ought to ask if the experimental techniques are over the boundary and amount to experimentation with people.
>
> Finally, you ought to ask yourself if the experiment *needs* to be done, rather than if it *ought* to be done, or if it *can* be done. If it is dangerous, or wrong, or both, and if it doesn't need to be done, just don't do it. This is not censorship. You must accept a physician's responsibility if, by free choice, you work within these classes of experiments.

Precedents exist for these attitudes, the memo reminded. Drug companies and medical investigators in this country must report their data to the FDA *whether or not* their technique or drug is eventually licensed for medical use. Hospitals are licensed by state medical boards. Federal grantees sign a paper foreswearing research on human subjects. Site visits precede approval of all research grants. Extensions of these precedents would be reasonable and desirable. All laboratories that work on animal virus genetics, on human cells, on human viruses, or on any tumor viruses should be licensed by the NIH–NCI (National Institutes of Health-National Cancer Institute), just as hospitals are licensed by medical boards. This licensing system should cover drug companies, foundations, universities, hospital research wings, and state and local health department laboratories. Peer group review of the physical state of a lab and of the overall competence of the investigators is now acceptable practice. Broad-based committees should be established with responsibility to hear scientists justify their line of experimentation to the public at large. It should be the burden of the investigators to demonstrate why they must do admittedly dangerous work. A progress report should be filed with the NIH a few times a year, and it should be made public. The memo concluded:

> No one should be permitted the freedom to do the first, most messy experiments in secret and present us all with a reprehensible and/or dangerous *fait accompli* at a press conference.

Science had attracted Robert Pollack originally because of what he from a distance thought was its priestly quality. And on becoming a scientist he did find himself in the midst of a kind of priesthood. Before he was through with four years at Columbia University, however, he learned to his dismay that many of the priests were concerned only with dogma to the exclusion of the welfare of their parishioners. He had in fact abandoned the study of nuclear physics because almost none of the physicists he knew cared enough about the social consequences of their work to debate positive and negative values among themselves, let alone with their nonscientist friends and neighbors. Yet Pollack was dedicated to the scientific method. He could not imagine any other occupation so satisfying as the one that privileged him to ask questions of nature and to know that he could depend on the answers. The scientist he had addressed by telephone on Monday afternoon, June 28, had a reputation for artfulness in phrasing the questions he asked of nature and for brilliance in interpreting the answers he got. Pollack's intuition told him that such a man could not escape caring about the effects of his sophisticated work on people less sophisticated than himself. But there was no way to tell for sure until Pollack heard from Professor Berg—if he ever would.

2

ROBERT POLLACK was worried about the dangerous potentialities of a hybrid virus molecule that had not yet been created. Dr. Andrew M. Lewis, Jr., a U.S. Public Health Service physician assigned to the National Institute of Allergy and Infectious Diseases at Bethesda, Maryland, was worried about the potential dangers of a family of five hybrids of SV40 that had already come into existence with man's unwitting connivance. Like Pollack, Dr. Lewis was a junior investigator, still in his thirties. Unlike Pollack, he had no cause to suspect that he might jeopardize his career by openly discussing his concern, and he prepared to do that at a Cold Spring Harbor Laboratory tumor virus meeting in August 1971, within a few weeks of Pollack's phone call to Professor Berg. Dr. Lewis had described one member of the worrisome virus family in the *Proceedings of the National Academy of Sciences* two years before. Accompanying the description had been this cautionary sentence: "Such viruses could be maintained in human and sub-human populations, and as pathogens would represent unknown hazards."

To Dr. Lewis's knowledge, no comparable declaration of possible danger from hybrid viruses had ever been published before anywhere in the literature. Yet no other experimenter had expressed interest in examining that hybrid until the spring of 1971, when Dr. Carel Mulder, one of Nobelist James Watson's senior research associates, asked for a seed from it. Dr. Lewis had agreed to provide the seed in September. In the meanwhile, he and his fellow workers at NIAID had discovered the four related hybrids of SV40, and he decided to make an informal report on them before fulfilling the September commitment.

Now science occasionally erupts virtually overnight with an entirely unexpected sensation, such as the discovery of the antibiotic power of bread mold. But most important scientific developments occur through a long, slow series of observations and experiments, whose sequence could not be foretold by anyone. The work Dr. Lewis was engaged in had progressed according to this more usual pattern.

During the war in Korea in the 1950s, anywhere from 25 to 75 percent of new recruits in the U.S. Army came down with grippe and flu-like infections soon after entering training camp. The usual treatments for such ailments proved ineffective. Military doctors assigned to the Walter Reed Hospital on the edge of Washington, D.C., were summoned to diagnose and deal with the strange maladies. By 1954 a virus previously unseen by the Army was isolated and identified as the cause of the epidemics. Immediately, this virus was matched against another, which doctors associated with one the National Institutes of Health had isolated the year before and identified as the causative agent of acute upper respiratory tract infections common to a great many children in the civilian population. The matching process disclosed that the sicknesses of the soldiers and the children were due to different strains of the same virus, which was therefore named Adenovirus because the strain that had been identified first took adenoids and tonsils as its principal targets.

In no time at all, the number of identified strains of adenovirus rose to seven. Following custom in medical research, these were numbered in the order of their discovery: Adeno 1, Adeno 2, Adeno 3, and so on. Orders were issued for vaccines to be made from six of the seven strains. Adeno 3 was exempt because, although it caused sore throats and infections of the pharynx, it left its worst marks on the eyelids in the form of conjunctivitis. Adenos 4 and 7 were judged the most suitable for military needs. Adeno 1, Adeno 2, and Adeno 5 were picked for inoculation of children in the general population, upward of half of whom, it turned out, carried in their blood antibodies capable of fighting off adenoviruses by the time the children were five years old—a sure sign that the virus had infected their adenoids and tonsils before that age.

Vaccines are made by growing virus in tissue culture. As the adenovirus vaccines were to be used in humans, cells of the human body would have been the preferred tissue for culturing. But in the 1950s the art of tissue culture was still in a primitive state, and the only human cells available for the purpose had been taken from cancer patients. To avoid risk of spreading cancer among humans, the adenoviruses were adapted to grow in nonhuman tissues. And they did grow quite successfully in cells taken from the kidneys of rhesus monkeys. Ten to a dozen successive generations of the already named six of the seven strains of adenovirus were grown in this manner. The ultimate progeny were then given to drug houses for use in the manufacture of vaccines. By that time it was 1956 or 1957. And by the end of the decade the vaccines were in widespread use.

The safety of all vaccines is an object of constant attention within the U.S. Department of Health, Education and Welfare. Various tests are employed in monitoring vaccine quality. One such test for the adeno vaccines involved the injection of fluid from uninfected rhesus monkey kidney cultures into hamsters. The result was that tumors grew in the hamsters. That was an awkward surprise, for there were no tumors in the rhesus monkeys. In an attempt to explain the mystery, more of the same fluid from the uninfected rhesus kidney tissue was injected into the kidney tissues of a different kind of monkey, the African green monkey (the rhesus is native to India). Instead of growing as they normally would, the African green monkey cells began to disintegrate. Something that could at first be identified only as a "vacuolating agent" was killing the cells. To find the killer, all the adenovirus vaccines were searched for a virus other than adenovirus. And every one of them was found to contain a monkey virus that had escaped notice before.

The same ban on human tissue culture that had caused the adenovirus vaccines to be grown in rhesus monkey kidney tissue had earlier caused the polio vaccines to be grown in kidney tissue of the same monkey. So the polio vaccines now had to be subjected to the scrutiny that the adenoviruses had just undergone. And their examination showed all the polio vaccines to contain the identical monkey virus that had been found in the adenovirus vaccines.

As had been the case with the adenoviruses, all monkey viruses discovered in the course of work on the polio vaccines had been numbered in the order of their isolation: SV (for simian virus) 1, SV2, SV3, etc. Up to the time contamination of the adenovirus vaccines had been detected, SV39 had been located. That made the name of the mysterious newcomer SV40.

The year of discovery of SV40 was 1960, a year of considerable agitation among public health officials. Tens of millions of people had been injected with the contaminated polio and adenovirus vaccines. There had been no sign of any unusual rise in the incidence of cancer among the vaccinated population. But the theoretical possibility was present. Watchful waiting was the only reasonable course to follow with vaccines of the past. However, measures had to be taken to protect future recipients of vaccine. The most obvious avoidance technique was to treat the adeno and polio vaccines with SV40 anti-serum, which should cancel out the SV40 effects. It worked that way for polio, but among the adeno vaccine family only Adeno 7 appeared to be freed of SV40. And when that seemingly purged strain of vaccine was injected into hamsters, it still produced tumors.

Remembering how DNA had been transferred from one species of pneumococcus to another in Fred Griffiths's laboratory in Great Britain in 1928, the NIH scientists suspected it might be happening again. They confirmed the suspicions by treating the Adeno 7 vaccine with Adeno 7 anti-serum and then injecting the vaccine into hamsters. This time no tumors grew in the hamsters. The DNA that had originally resided in the SV40 had combined with the DNA in the Adeno 7. A hybrid virus had been created.

All this had happened before Dr. Lewis left Duke University Medical School to join the NIAID staff in 1963. Work on the Adeno-

SV40 research project was continuing at the time of his arrival at Bethesda, however, and within three years he was part of it. The investigative unit he belonged to concentrated on progeny of the original Adeno 2 strain of virus. These all turned out to be hybrids—defective hybrids. That is, they needed the assistance of another virus (what virologists call a "helper") in order to infect humans or other animals. The chance that they alone could possibly be responsible for a human epidemic was extremely small. But as successive generations were bred, nondefective offspring began to appear. The first to be identified was the one described in the *Proceedings of the National Academy of Sciences* in 1969. It clearly possessed infective power, and was named Adeno 2 + ND_1.

Once a new organism has been described in the scientific literature, the usual courtesy is for the person who describes it to make seed from it available on request. Dr. Carel Mulder's request had been granted in keeping with that tradition. But the people at NIAID were not entirely agreed that in this particular instance the seed should be released without some form of control over its distribution. In any case, September had not yet come when Dr. Lewis and an NIAID associate, Dr. Arthur Levine, arrived at Cold Spring Harbor to report that Adeno 2 + ND_1 had four close nondefective relatives: Adeno 2 + ND_2, Adeno 2 + ND_{3A}, Adeno 2 + ND_{3B}, and Adeno 2 + ND_4. Descriptions of these had not yet been published and the usual rules of scientific communication said that until publication did occur the reporting of data was respected as preliminary and confidential and the reporters of the data were not obliged to release their experimental materials for corroborative experiments by others.

Drs. Lewis and Levine read separate papers in tandem on behalf of themselves and five other NIAID colleagues at an evening meeting. After they had finished, a coffee break was announced. The coffee was served in a small room adjoining the meeting hall. At the end of the break, Dr. Lewis was confronted by a tall, gangling man whose head was wreathed by runaway wisps of red hair. Apparently not thinking it necessary to identify himself, the redhead barked: "You have no right to be up here talking about these hybrids unless you'll send them up here so we can study them." Dr. Lewis could not recall ever having been addressed so rudely before. His antagonist's voice carried the ring of authority, but who did the voice belong to? Could he be the notoriously fiery-tempered James Watson, the "Honest Jim" of *The Double Helix,* winner of the Nobel Prize? It was, and Dr. Lewis next heard himself being accused of conspiring with other scientists at NIAID to keep the nondefective Adeno-SV40 hybrids away from scientists employed outside the government. The prospect of any one or all of three punishments was held before him:

1. A letter, written by Watson to the director of the National Institutes of Health, charging Dr. Lewis with participation in a conspiracy to keep important scientific information from scientists outside the government, or

2. A letter from Watson to the editor of *Science,* journal of the American Assocation for the Advancement of Science, making the same allegations, or

3. A letter from Watson to the chairman of a key committee in Congress charging Dr. Lewis with improper use of public monies.

Dr. Lewis could understand why the head of the Cold Spring Harbor lab would be eager to get hold of the five nondefective Adeno-SV40 hybrids. Very large sums of money had been freed by President Richard Nixon's impetuous campaign to abolish cancer within an impossibly short span of time. Competition for those funds had been fierce. Nobelist Watson held a $5 million-a-year grant from the National Cancer Institute, and was under great pressure to seize every likely research opportunity, for good approaches to the cancer problem are always scarce. The Adeno-SV40 hybrids held unusual promise because each of them contained a different segment of the SV40 DNA. Mere comparison of their effects would be bound to turn up at least some fragmentary hints on how the SV40 tumor control gene did its work. Even during the Cold Spring Harbor meeting, the competitive tension could be felt to tighten when Dr. Daniel Nathans, of Johns Hopkins University Medical School, announced that he had begun to use a newly discovered enzyme, manufactured by an influenza bacterium, to chop SV40 DNA into eleven pieces of different sizes in order to study the function of each piece and the order in which the functions were expressed.

The conspiracy charge Dr. Lewis could not understand. It had no basis whatever in fact. He welcomed the end of the half hour of confrontation with Watson, for he needed time to clear his head and to give thought to the six NIAID men who were working with him on the nondefective hybrids. All of them were young. All were just starting on their research careers. All would be defenseless against public chastisement by someone of Nobel stature.

After listening to the other papers on the tumor virus meeting program that night, Dr. Lewis sought out older scientists he trusted and discussed the Watson encounter with them. Although they split about equally on the question of how dangerous the Adeno-SV40 hybrids might be, they agreed unanimously that Dr. Lewis had no obligation to release the four newly described hybrids to anyone until the data on them were appropriately published.

Although he had told his accuser that any of the three suggested penalties would be difficult to resist, Dr. Lewis went home to Bethesda determined to resist them if they arose, which they never did. Dr. Lewis also decided, as Robert Pollack had done earlier, that as a scientist he had a personal responsibility that could not be blinked away. So far as he could see, the five Adeno-SV40 nondefectives were genuinely risky. Although he had never completed the specialty board examinations that would allow him to practice as such, he had been trained as a pediatrician. He had two young children of his own. He would not choose to expose them to adenoviruses embodying the tumor-provoking potential of SV40. He assumed that his concern for their welfare was no greater than the concern of other parents for their children would be. When asked about his decision later, he explained:

> The essential problem was that human adenoviruses infect young children. Fifty to ninety percent of the kids in the general population have antibody by the time they are five years old. Not only do they have antibody, but Adeno 2 establishes a chronic infection in the tonsils and adenoids. This virus sets up a chronic infection of lymphoid tissue. To risk establishing in the population at large, and especially in young children, an agent that contained the region of the SV40 genome that seemed to be associated with malignant disease just didn't make sense.
>
> Well, how do you keep from doing that? Adeno 2 is not a highly infective virus. If you slop a little bit of it on the floor, you're not going to get infected with it. In the laboratory, it is a difficult agent to infect people with. The NIH scientist who originally isolated the virus back in the late 1950s really had to work hard to infect people with Adeno 2. They finally had to swab it on the conjunctival with a cotton-tipped applicator in order to induce clinically apparent disease, and that took a large dose of virus. Still, we didn't know what would happen if somebody would track small amounts of the virus home. Kids could get infected, and they could sneeze on the other kids, who could get infected. This could be an endemic type of infection. It wouldn't be an epidemic type of thing like influenza, where 40 or 50 people would become acutely ill. But no matter what the infection was, we didn't want to risk seeding the virus into the population if we could help it.

Accordingly, on September 1, the agreed day, Dr. Lewis sent to Dr. Carel Mulder at Cold Spring Harbor seed of the Adeno 2 prototype strain and the Adeno 2 + ND_1 hybrid, along with a letter identifying the type of cell culture in which the hybrid had been grown and recommending that the type be used exclusively in future culturing. The letter said:

I would also recommend that your stocks of ND_1 be prepared in a freshly sterilized cubicle, or, preferably, an area in which SV40 has never been handled. As I am sure you are aware, working with both ND_1 and SV40 in the same area without extensive precautions to avoid cross-contamination is asking for trouble.

We recommend that you consider the ND_1 virus a potential biohazard. In the interest of the public safety, before large scale studies are undertaken, we urge that you screen all personnel who will be working with this agent for the presence of Adeno 2 neutralizing antibody. Only those with evidence of such antibody should be allowed to handle the agent. During such studies, all areas and apparatus used to handle the ND_1 virus should be cleared immediately after use with an oxidizing agent such as sodium hypochlorite or tincture of iodine; we also use ultraviolet lights at night in our laboratory area. All children, especially those under five years old, should be permanently excluded from the area where this virus is being used. Personnel working with this agent who develop pharyngitis, conjunctivitis, or viral pneumonia, should be screened for infection with ND_1 virus both by virus isolation and by serologic testing for Adeno 2 antibodies.

Though no data to date indicate that the ND_1 virus is, indeed, a dangerous agent, I am genuinely concerned about the likelihood of this virus spreading through the general population.

A telephone conversation concerning that letter took place on September 10, and five days after that Dr. Lewis again addressed Dr. Mulder, as follows:

"After further discussion about the safety precautions we should follow in distributing the Adeno 2 + ND_1 hybrid, I would like you to consider not supplying virus seed to other laboratories. If you receive such requests, I ask that you refer those individuals to me . . . [but only] to be sure that all workers who are propagating this agent are informed of the safety problems and are acting responsibly in handling the virus."

By that time, twenty or more other researchers in different parts of the world had asked for seed of Adeno 2 + ND_1. Dr. Lewis sought from each of them the same guarantees that he had requested of Dr. Mulder. Several refused. To comply, they said, would establish an undesirable precedent in restricting science's traditional freedom of inquiry.

3 ROBERT POLLACK treated his telephone talk with Professor Berg as a private matter. One man's conscience had spoken to another man's conscience. No word of it reached the eyes or ears of the general public for four years. But news of it began buzzing over the internal communications system of the scientific community almost as soon as Pollack and Berg hung up their respective telephones. And the supposedly confidential system developed a momentary leak in September 1971. The exact date is uncertain, but sometime during the month Spyros Andreopoulos, director of public information at Stanford University Medical School, got a phone call from the University of Illinois. Rick Polk, a reporter on the student newspaper, was on the line. Polk said that an Illinois microbiology professor had made some remarks to a group of students about genetics research and hazards related to it. In essence, Polk said, the professor took exception to an experiment that Berg was planning. What it amounted to, as Polk understood it, was that Berg was manipulating DNA molecules to produce a cell-replicated carcinogen, that is, a cancer-causing agent. Polk asked Andreopoulos for an informational readout on Berg in general and on the described experiment in particular.

Andreopoulos told Berg about Rick Polk's query. Berg called the Illinois professor and explained the experiment to him. The professor said that his remarks "had been misunderstood," and that the professor would talk to Polk about it. Andreopoulos heard nothing further from the Illinois student.

4 DURING THAT same month of September 1971, young Dr. Andrew M. Lewis, Jr., had an unexpected visit from a man he had never met before. Professor Paul Berg dropped into the NIAID laboratory at Bethesda to talk about the SV40-lambda phage-*E. coli* hybrid experiment then being assembled at Stanford. The professor explained that other scientists were questioning whether it might not be too risky to try to insert the hybrid DNA molecule into the *E. coli* bacterium because of *E. coli*'s habitation of the human gut. The parallel between that risk and the risk of establishing Adeno-SV40 hybrid viruses in the upper respiratory tracts of very young children was too clear to pass over, and the

two men spent some time discussing what amounted to a mutual problem.

Professor Berg's visit to NIAID was only one of many that he paid to fellow scientists during the last quarter of 1971 and the first quarter of 1972. The subject of the conversation was always the same: Was the proposed SV40-lambda phage-*E. coli* experiment too dangerous to undertake?

At first, the professor had no doubt of the correctness of his original reluctance to take Robert Pollack's worries seriously. As a veteran nonresident fellow of the Salk Institute, Professor Berg was thoroughly familiar with the record of contamination of polio vaccines by SV40. Absence from that record of any apparent abnormal trend in malignancy among the vaccinated population had reassured him. Further confidence had been generated by Nobel laureate virologist Renato Dulbecco's blithely repeated offer (for which no one had ever called him to account) to drink SV40 to prove its harmlessness to man. But the more colleagues Professor Berg consulted, the less sure he became of his original assessment of the risks.

At Massachusetts Institute of Technology, for example, Dr. David Baltimore, holder of a Nobel Prize for his share in the discovery of reverse transcriptase—the enzyme that enables RNA, the nucleic acid that normally copies the hereditary instructions from DNA, to copy the message back into DNA—voted "no" when Berg invited his opinion. At the National Cancer Institute at Bethesda, Dr. Maxine Singer was very critical of the experiment in spite of her warm personal friendship with Berg. Edward Ziff, a young American student of genetics working at Cambridge University in England, also considered the hybrid manipulation a dubious proposition. "The Berg experiment scares the pants off a lot of people, including him," Dr. Wallace Rowe, a National Institutes of Health virologist, was quoted as saying, in *Science,* journal of the American Association for the Advancement of Science. Another specialist in virology, Dr. George Todaro, of the NCI, said the experiment was "one of those I think just shouldn't be done." And Berg heard still more unfavorable comment in Europe, particularly at an all-night beer party on the roof of an old castle overlooking the Straits of Sicily.

The party was a side effect of an annual molecular biology workshop sponsored by the North Atlantic Treaty Organization. Professor Berg was a guest lecturer at the workshop. His subject, naturally, was the SV40-lambda phage-*E. coli* hybrid. The young Europeans in the audience reacted volubly. With the ghost of Adolf Hitler still goose-stepping among them, they saw themselves poised on the brink of a frightening new era of human engineering and behavior control. After the formal lecture period, they asked for an informal discussion of political and social consequences of the hybridizing experiments. In response to that request, lights were strung on the top of a castle prominent on the Sicilian shoreline. About eighty people came, drank beer, and argued until midnight: "Should we do this or shouldn't we do it?" Ethical and moral considerations had the floor without interruption. That approach did not appeal to Berg. He would have preferred to discuss the problem in terms of health hazards. But he couldn't get a word in edgewise.

The behavior of the most eminent person present on the castle top—Sir Francis Crick—fascinated Berg. Crick didn't say a word all evening. When Berg personally asked the knighted Nobelist for an opinion, he could elicit none. It seemed exceedingly odd to Berg that a man who knew as much about DNA as Crick did, a man who has written a book on the philosophy of science, would not feel compelled to say something on such a critical occasion.

After returning home to California, Professor Berg assessed the cumulative effect of six months of self-examination. He finally brought himself to face the fact that although the probability of danger in the SV40-lambda phage-*E. coli* experiment seemed low to him, he couldn't honestly argue that it was zero. Certainly it was finite. He recognized that it was in the nature of science to err. He had been wrong many times in predicting the outcome of experiments in the past. Should he be proved wrong this time, the consequences wouldn't be such that he would want to live with them. Were he alone, he would risk it. But he was not alone.

Having made his decision, he hunted up his principal collaborator, Dr. David Jackson (who was soon to leave for the University of Michigan), and obtained Jackson's agreement to drop that part of the experiment that would put the SV40-lambda phage hybrid into *E. coli.* Then Professor Berg returned the phone call that Robert Pollack had made to him six months earlier.

"We are not going to do that part of the SV40 hybrid experiment that worried you," Berg told Pollack. "You were right about it. We had not fully considered the possibly hazardous consequences."

Pollack was elated, and expressed himself accordingly.

But Berg had more to say: "It seems plain to me that we are only beginning this type of research. I feel it is important for us to get a fix on just what kinds of hazards tumor viruses themselves pose to man. Since you feel so strongly about this question, will you help to organize an international conference on the subject for early next year?"

Pollack gladly accepted the invitation.

5

BEFORE ABANDONING his plan to grow a hybrid SV40-lambda phage DNA molecule in the *E. coli* bacterium, Professor Berg paused to stake out a formal claim to priority on a design for hybridizing, in the laboratory, DNA molecules of any two living species. The claim took the form of a report to the *Proceedings of the National Academy of Sciences.* Berg submitted it on July 31, 1972 in his own name and the names of two postdoctoral associates, Drs. David Jackson and Robert Symonds. It was an interesting document, at least as much for what it suggested as for what it said outright, and it deserves more attention than it has received, either in the scientific journals or in the popular press.

One of the footnotes to the paper discloses that the gene-splicing experiment conducted by the Jackson-Symonds-Berg team (to put the authors in the correct order of the amount of laboratory work each did) was simultaneously paralleled by another experiment with fundamentally identical objectives. One of two participants in that experiment Berg identified as a fellow member of the National Academy of Sciences and of the Stanford Medical School faculty, Professor A. Dale Kaiser. The second participant was Peter Lobban, a graduate student who reached Stanford from Maryland by way of Massachusetts Institute of Technology in 1966.

Lobban was named five times in the Jackson-Symonds-Berg report to *PNAS,* thanked "for many helpful discussions," and specifically credited with prior demonstration of the necessity for using certain enzymes that proved crucial to the success of the Berg team's experiment. It is therefore curious that Lobban should be almost totally neglected, as he has been, in popular accounts of the origin and development of one of the most pivotal, if not the most pivotal, series of scientific discoveries in the history of homo sapiens.

Lobban was, as I have said, a Stanford graduate student. Professor Kaiser was his Ph.D. adviser. As background for the story of Lobban, then, we ought to know something about Kaiser.

Kaiser had spent most of his professional life in the "phage school" of genetics. He had done landmark work on the lambda phage that Dr. Esther Lederberg had found attached to the K12 strain of *E. coli.* The lambda phage, it will be recalled, has the fascinating option of killing or living peaceably as a temporary in-law and guest of *E. coli.* What makes it possible for the lambda to behave in this remarkable way?

Kaiser, among many others, had pursued that tantalizing question.

The answer begins with geometry. On entering *E. coli* K12, the lambda assumes a circular shape and rolls along the bacterial chromosome until a particular sequence of genetic alphabet bases (As, Cs, Gs, and Ts) on the chromosome matches a sequence of bases on the lambda's own DNA. The sequence is fifteen letters long and spells out this genetic latch: GCTTTTTTATACTAA
CGAAAAAATATGATT.

A rather large parking sign for such a small organism as the lambda phage? One must think like a lambda phage to answer correctly. Kaiser, who is by now expert at anticipating the perceptions of microorganisms, says that even a string of fifteen genetic alphabet letters doesn't stand out at all sharply on the relatively enormous length of the bacterial chromosome. To the incoming lambda, the chromosome looks as bewildering as an international airport would look if the whole airport were filled with solid rows of automobiles and you were looking for your car somewhere among them.

To guide the lambda, nature has marked the *E. coli* chromosome with about fifty more genetic alphabet letters on either side of the fifteen-letter-long lambda stop sign to make sure the phage can find its way, just as a thoughtful parking lot attendant might put a long array of distinctive pennants around your car for your convenience. With that help, the lambda finds the spot where it belongs on the *E. coli* chromosome, stops there, and unrolls into the linearity of the chromosome as shown below.

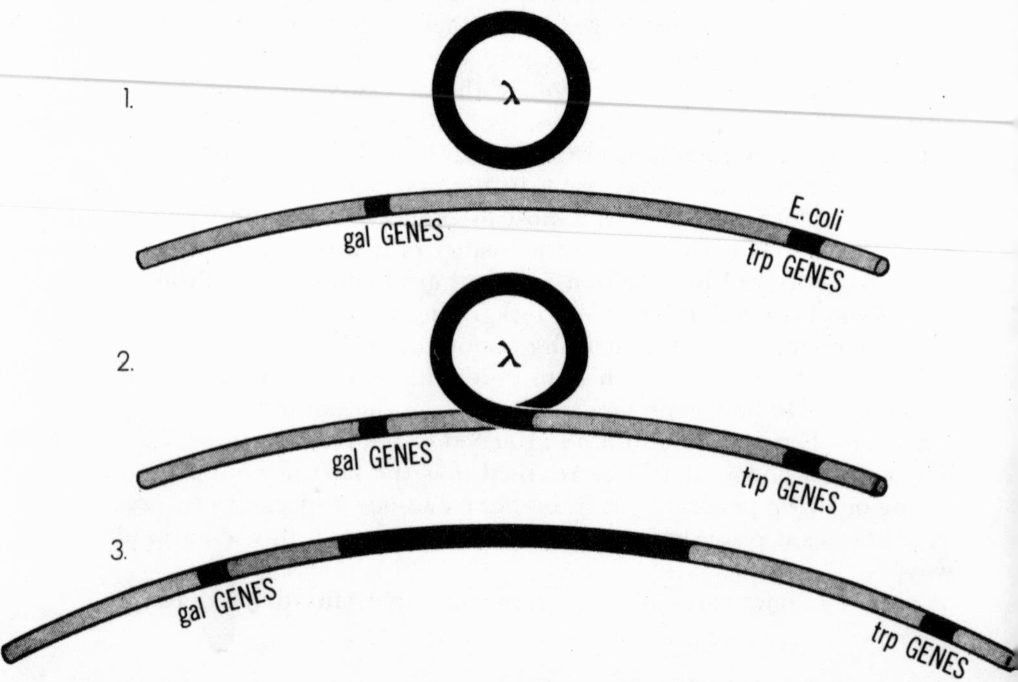

But how does the lambda DNA change from the linear form in which it enters the *E. coli* to the circular form that makes the next step possible inside the *E. coli*? The answer is that the ends of the linear piece of DNA are "sticky." Instead of coming out even, as they do in normal DNA molecules, the ends of the lambda strands are out of register. One strand projects at one end of a molecule, while the other strand projects at the opposite end as shown here:

NORMAL DNA MOLECULE

These "sticky" projections enable the ends of the separated strands of DNA to find and grab hold of each other, so:

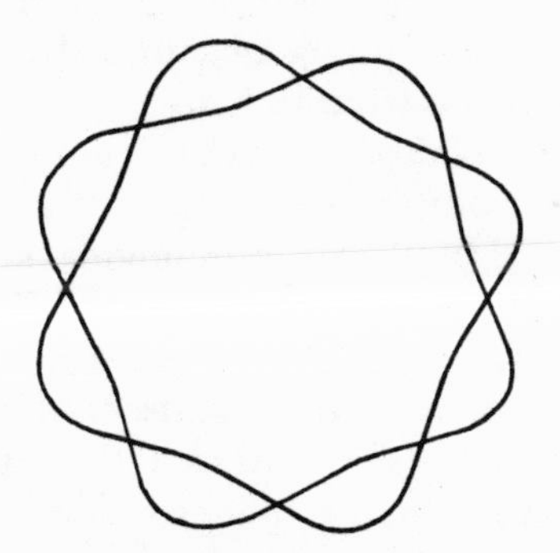

lambda DNA molecule

What makes the "stickiness" adhere? One of Kaiser's postgraduate associates, Hans Strack, along with a graduate student named Ray Wu and Kaiser himself, investigated that conundrum. After three years of work they discovered a specific sequence of twelve genetic alphabet letters on one end of a strand, like this: GGGCG-

GCGACCT, and on the other end of that strand a complementary sequence like this: CCCGCCGCTGGA. As always, the As were opposite Ts, the Cs opposite Gs.

Later pieces of scientific detective work in which Professor Kaiser and his staff participated finally uncovered the fact that the lambda phage has about fifty genes, forty-eight of which are somehow turned off during the phage's peaceable residence in th *E. coli* chromosome. Of the remaining two genes (if indeed there are two), one is the "suppressor, or "off" switch.

History did not take Kaiser by the hand in 1960 and say, "come this way." With no premonition of any kind, he was, as usual, following his highly inquisitive nose when, in collaboration with Stanford Professor David Hogness, he did the experiment that would lead to a simple method of splicing genes twelve years later. The objective in 1960 was to smuggle lambda DNA into the *E. coli* bacterium to propagate there. The trick had been turned with other bacteria, but not with *E. coli.* One inorganic substance after another was added to the nutrients in the *E. coli* cultures to see if they would serve as burglars' tools. Calcium and magnesium did the job, and did it better together than separately. But neither way was very efficient. So the lambda phage itself was called in to help. This time the DNA cargo went through along with the lambda. Further experimentation showed that the DNA transfer could be made with the help of any phage whose DNA strands had "sticky" ends to match the "sticky" ends of the DNA that was being moved. To this day no one understands why the calcium and magnesium are better for the purpose or how the "helper" phage accomplishes its mission. All that is known for sure is that the trouble Kaiser and Hogness encountered in 1960 lay in the DNA's inability to survive after it got inside the *E. coli.* The situation was comparable to that of a transplanted human heart, which is perceived by the immunological system of its new host as a threatening stranger who must be mercilessly attacked.

Six.years after completion of the first successful experiment in that series, Lobban came to Stanford. Kaiser rated him among the brightest students who had ever entered the Biochemistry Department. Lobban's imagination, in the professor's opinion, was "fantastically good." Moreover, it was a highly practical imagination, in some ways more characteristic of engineering than of science. It was possibly a manifestation of inheritance, for Lobban's father was a chemical engineer. Kaiser, therefore, was not surprised when Lobban perceived the 1960 experiment as the start of a major breakthrough and took it as the jump-off point for a much more daring scheme: to breed a versatile stable of phage for the purpose of carrying DNA *on order* from *any* species of life to *any other* species regardless of normal and natural relationships.

Lobban did not claim any new discoveries. He did not offer any new inventions. He simply put together a series of repeatedly confirmed observations and repeatedly tested laboratory techniques in a sequence that no one else, so far as he or Kaiser knew, had ever hit upon. In fact, the concept was so different from the then accustomed ways of seeing things that Lobban himself felt a little shaky about it at first. To check himself out, he orally ran the idea past several Stanford scientists who had successfully used individual components of the system he had in mind. They all discouraged him. What he proposed to do, they said, just would not work.

But Lobban's own genes carried a strong streak of tenacity (as well as a sober, almost prim mien) from his Scottish ancestors. The potential future of his scheme, he decided, was too promising not to run the risk of failure. And on November 9, 1969, Lobban presented to a committee of three Stanford professors a typed project proposal twelve pages long, with three additional pages of literature citations, in fulfillment of a preliminary requirement for the Ph.D. degree.

"Transducing bacteriophage . . . have proved to be an excellent source of small, genetically well-defined portions of the DNA of their hosts [that is, of the bacteria in which the phage had taken up lodging]," the proposal began; the usefulness of those DNA portions had "been amply demonstrated" in studies of how gene action in *E. coli* is started, controlled, and stopped. "At present, however, the usefulness of transducing phage is limited" to moving genes that are located near the sites on the bacterial chromosomes where the phage attach themselves. It would be desirable to create new transducing phage containing any desired genes, including those from organisms other than the host of the phage. The desirable was possible, thanks to the "large body of knowledge concerning lambda phage and its host, *E. coli*" and "recent advances in enzymology." A convenient method would be "linking together appropriate DNA pieces," that is, constructing hybrids and "bringing those genomes into *E. coli* cells, where they can be handled like naturally occurring transducing phage."

To soften the startling effect of his proposal, Lobban reminded his Ph.D. judges of the Kaiser-Hogness experiment of 1960. Then he recalled what Kaiser and his students and scientist associates had learned about the "sticky" ends of lambda phage DNA.

One confidently could, he said, break open the lambda circle, insert chosen foreign genes in the cleavage, and close the circle again if proper care were taken in advance to set up the right combination of the nitrogenous bases of the genetic alphabet (As, Cs, Gs, and Ts) to make the augmented ring cohere securely. He specifically named five enzymes known to be doing various aspects of such work—some that chewed up DNA, some that added onto DNA, and some that glued DNA together and filled in small nicks. Bacteria had been observed to use those enzymes to repair their own DNA or to chop up and destroy the DNA of invading viruses.

Finally, Lobban listed nine steps of laboratory procedure, which he said would accomplish the objectives of his proposal, and offered two rough diagrams of the result he expected. One of the diagrams is reproduced here.

"Assuming its success at simpler tasks," Lobban proposed an eventual goal for his method: "to produce a collection of tranductants synthesizing the products of genes of higher organisms." In plain English, to transport eukaryote DNA into *E. coli* by bacteriophage and there use the prokaryote bacterium's reproductive capabilities to mass-produce the products (enzymes and proteins) ordered by the eukaryote genes.

Stanford's Ph.D. examinations are notoriously tougher than those of most other schools. Members of examining committees are exhorted to be rigorous in exploiting holes in the logic of degree applicants. Lobban's proposal took a battering. But no weaknesses were found in it. Nor did any of the professors question its originality. And none expressed awareness of anything like it.

To start his project, Lobban chose the temperate virus P22, which attacks the *Salmonella* bacillus. It is a simple virus, and grows rapidly. He anticipated no serious difficulty in his first operation, which was simply to break open with detergent two P22 molecules, extract the DNA from them, open the DNA circles, and splice them together to form one circle twice as big.

A problem did arise, however—one he had no reason to expect. Another scientist in the Stanford Biochemistry Department brought him the news that an experiment very similar if not identical to his was underway in the laboratory of the department chairman, Professor Berg.

None of the members of Lobban's examining committee had known of Berg's activity. Lobban was chagrined. As a graduate student, working alone in the Stanford tradition, he would have little chance in competition with a professor with a formidable reputation and two experienced postdoctoral researchers even if he enjoyed that sort of brain-racing, which everyone had read about in *The Double Helix.* Lobban intuitively preferred the conduct of Charles Darwin when Darwin, in the midst of his opus on evolution, had received a brief but essentially identical statement of evolutionary theory from

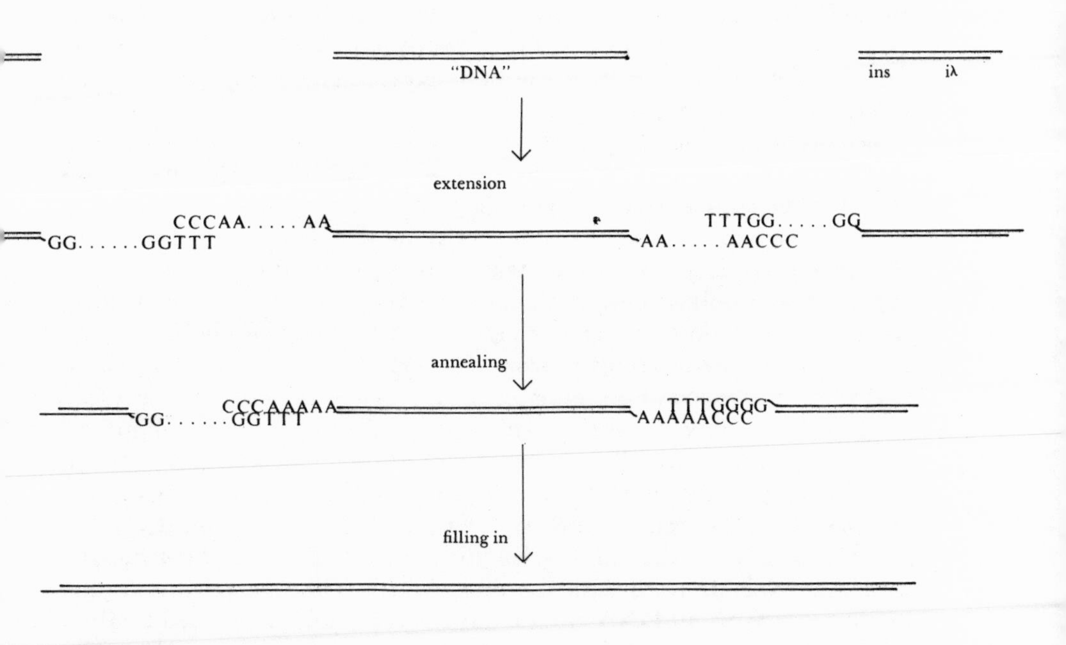

The double-stranded segments on the right and left sides of each line of the diagram marked "A" and "ins," represent the ends of the severed circle of lambda phage DNA. The segment in the middle of each line, marked "DNA," represents the galactose genes from *E. coli.* The second line of the diagram shows how appropriate sequences of DNA subunits are added to match As with Ts and Cs with Gs. The third line shows how these extensions are pulled together without disturbing the order of the letters of the genetic alphabet. The bottom line shows the solidly constructed hybrid molecule, "sticky" (that is, out of register) at both ends and therefore ready to close itself into a circle.

Alfred Wallace, half the earth away. Darwin had published Wallace's paper as a companion to his own. Lobban admired that. And he knew that Kaiser did too. After deciding that the gentlemanly precedent was the one to follow, Lobban had to choose one of Berg's postdoctoral associates and offer to share data with him. The choice was Jackson, who confirmed the Berg experiment, accepted Lobban's offer, and agreed to reciprocate.

Because P22 is not only easier to grow but easier to manage than is SV40, Lobban completed the joining of the two P22 DNA molecules ahead of the similar step in the Berg camp, and in the course of doing so discovered that a lambda phage enzyme that chopped off the ends of DNA strands in a special way was vital to the success of the experiment. Lobban passed that information on, and, in fact, gave the Berg team some of that enzyme to work with. The help Lobban got in return came mostly from Robert Symonds, an Australian wizard in radioactive tagging of DNA, without which it is often impossible to tell what is happening in DNA splicing experiments.

Lobban had refrained from what would have been entirely ethical individual publication of his findings at the time he made them. He assumed that his report and the Berg team's report would appear simultaneously in the same scientific journal. But by the time the Berg team was finished with the first two steps of its experiment—the joining of two SV40 DNA circles and the injection of coupled lambda phage and *E. coli* genes into an SV40 circle—Berg was under heavy pressure first from Robert Pollack and then from many other scientific critics of the original intention to grow the SV40-lambda DNA hybrid in *E. coli.* Berg went to see Kaiser and explained why immediate publication would be most desirable. Kaiser talked to Lobban about it. Lobban felt he could finish his report in time to meet Berg's deadline. But Kaiser, fearful that too much cramming would hurt the quality of Lobban's Ph.D. thesis, urged his student to take the conservative course and wait for later recognition. Lobban accepted Kaiser's advice and has never since complained that he did so, although some of his friends point out that lack of appropriate credit deprived him of top standing in the job market when he left school and finally removed him from the area of research that he had helped to pioneer.

Berg submitted the Jackson-Symonds-Berg report to the *PNAS* between two and three months after Lobban turned in his finished thesis. Because of the time pressure, many details of the Berg team's experimental procedure were omitted from the *PNAS* and promised for publication elsewhere. But Berg, as a professor and department chairman, could afford to be bolder in stating his ultimate purpose than Lobban dared to be. As befitted a student, Lobban had said that exploration of animal cells would be attempted *if* the DNA molecule splicing method succeeded in simpler tasks. Berg spoke directly to the main point.

"Our goal is to develop a method by which new, functionally defined segments of genetic information can be introduced into mammalian cells. . . . Accordingly, we have developed biochemical techniques that are generally applicable for joining covalently any two DNA molecules. . . . Such hybrid(s) . . . and others like them can be tested for their capacity to transduce foreign DNA sequences into mammalian cells, and can be used to determine whether these new nonviral genes can be expressed in a novel environment."

The *PNAS* report said the first step in execution of the Berg team's experiment was the opening of the closed circles of SV40 and lambda phage DNA in order to get linear molecules with free ends. This was done by bursting the molecules with detergent and separating out the DNA by chemical sedimentation, then mixing the DNA circles with a "restriction" enzyme, which severed the circles as cleanly as a surgeon's scalpel.

Bacteria secrete restriction enzymes to protect themselves against invasion by viruses. The enzymes are released automatically to seek out and chop up viral DNA as it comes through the bacterial cell wall. The enzymes work just as effectively against the DNA of the bacteria themselves. So the bacteria, in fighting the viruses, protect their own DNA from damage just as policemen wear bulletproof vests: the bacteria cover their DNA with enzyme-resistant chemicals through a process known as methylation.

The earliest isolation of a restriction enzyme had been accomplished in 1968 by Professor Matthew Meselson of Harvard. But no practical use for it had been seen. That limitation did not apply to subsequently discovered enzymes. It will be remembered, for example, that Professor Daniel Nathans, of Johns Hopkins University Medical School, had produced a mild sensation by announcing, at the Cold Spring Harbor Laboratory in September 1971, that he was using a restriction enzyme from an influenza bacillus to cut SV40 into eleven pieces of various sizes. And greater fame was to come to another restriction enzyme identified in 1969 by Robert Yoshimori, a graduate student in the laboratory of Professor Herbert Boyer at the University of California (San Francisco) Medical School. Yoshimori called his enzyme EcoRI, indicating that it had come from a strain of *E. coli* (Eco) that resisted (R) antibiotics. Boyer gave a small supply of EcoRI to Berg for experimental purposes around the time Yoshimori's discovery was announced in 1971. And Berg used EcoRI in designing his hybrid DNA molecule.

In addition to EcoRI the Berg team used several other types of enzymes, one of which is made both by bacteria and by viruses. It repairs nicks and breaks in the gene sequences, even to the extent of adding or excising whole chunks of genetic information at strategic points when occasion requires. It is called ligase. It was discovered independently in five different laboratories in 1967. In 1970 a research group headed by H. Govind Khorana, who was then at the University of Wisconsin, slowly putting together the first synthetic gene (for the completion of which he later received a Nobel Prize), found that a ligase produced by a bacteriophage named T4 could sometimes join together entirely severed ends of double-stranded DNA fragments. For an efficient jointure, there had to be a mechanism to hold the ends together long enough for the ligase to act. Khorana's solution was to hunt for severed ends of double-stranded DNA that had single-stranded tails.

One way to guarantee single-stranded tails was to expose a double strand of DNA to an enzyme produced by the lambda phage. Lambda exonuclease, it was called. "Exo" means "at the end," and this enzyme chewed off one end of one strand of DNA and the other end of the opposite strand. It was the enzyme Lobban found essential to gene splicing.

Once the exonuclease had completed its work, it was possible to call upon still another enzyme—terminal transferase—to add a series of identical chemical bases of the genetic alphabet (T T T T T T in this case), one by one, to the single-stranded tail of one DNA segment and a complementary series of bases (A A A A A A in this case), one by one, to the tail of the other DNA segment. The genetic alphabet blocks acted as interlocking teeth to hold the hybrid molecule together until DNA ligase could be added to seal remaining chinks in the structure.

Now all construction projects require dependable inputs of energy. Nature packages energy in two elusive chemicals—ATP and TTP—that dart about our bodies like will-o'-the-wisps for immediate use when and where needed. The Berg team added small amounts of them to the gene-splicing formula, and they supplied ample power for the job.

The project was titled "Biochemical Method for Inserting New Genetic Information into DNA of Simian Virus 40." A flow chart from the Berg team report appeared in *PNAS* in October 1972. Granting a difference in the sophistication of the drawings, its similarity to Lobban's Ph.D. project proposal sketches was evident.

No history of the underlying idea was offered in the *PNAS* text. In the subsequent interview that was mentioned earlier, Berg said his interest in mammalian cells had been aroused originally by conversations with Nobelist Renato Dulbecco at the Salk Institute for

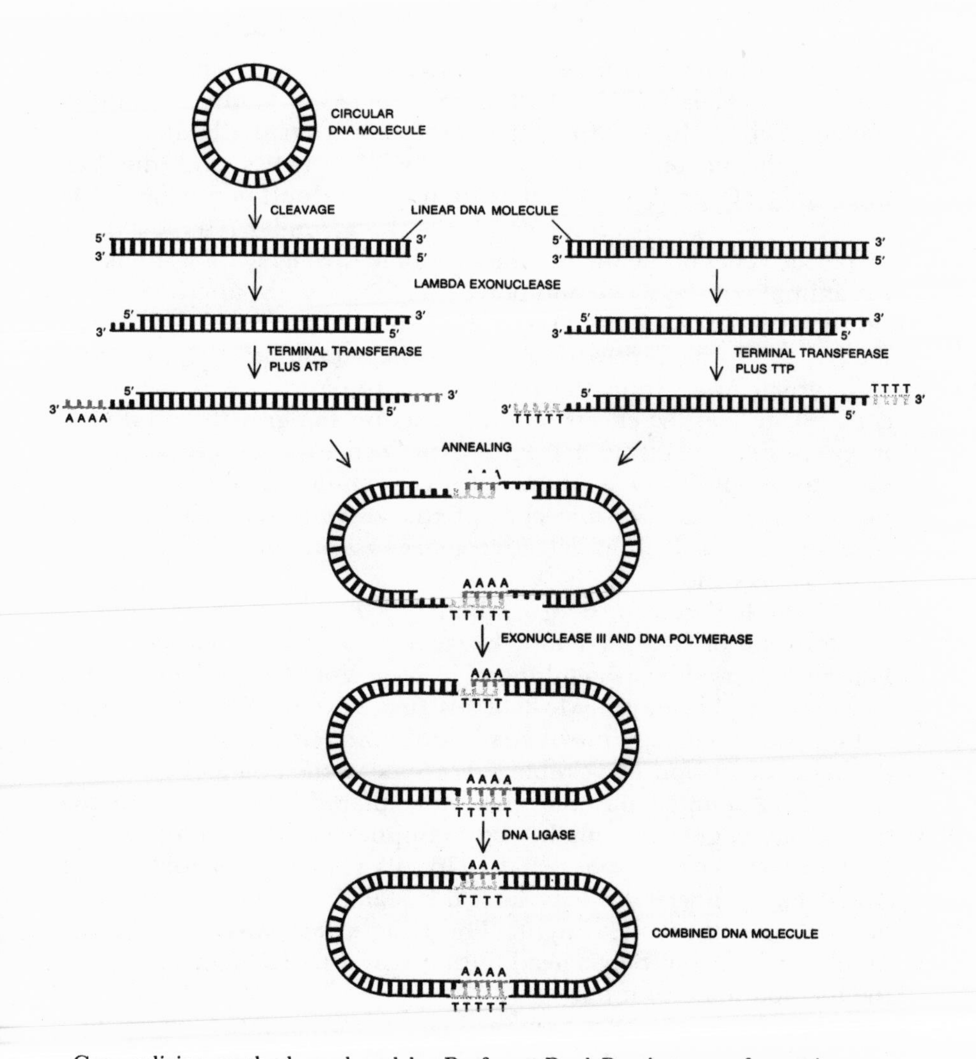

Gene-splicing method employed by Professor Paul Berg's team of experimenters at Stanford began with cleavage of circular DNA molecule *(top left)*. The resulting piece of linear DNA was then treated with an enzyme that shortened one end of one strand of the DNA and the opposite end of the second strand. The foreign DNA *(top right)* that was to be spliced into the original DNA molecule was meanwhile treated by the same enzyme with the same result. A clump of nitrogenous bases corresponding to the genetic alphabet letter T were then added to the foreign DNA *(third step down, right),* while bases corresponding to the complementary letter A were added to the original DNA *(third step down, left)*. When the original molecule was closed up with the foreign DNA as an internal patch, the As and the Ts hooked on to each other and required only the addition of a sealing enzyme to complete the job. *From "The Manipulation of Genes" by Stanley N. Cohen. Copyright © 1975 by Scientific American, Inc. All rights reserved.*

Biological Studies in 1965 and then triggered into action by a lecture with which Kaiser had closed a new course of study at Stanford, Biology 211, in June 1967. After reviewing what the lambda phage had taught its observers about bacterial genetics and the basic mechanics of heredity, Kaiser in that lecture outlined what he felt were favorable prospects for accumulating a similar reservoir of lore about the genes of the mammalian genome through explorations with the animal viruses, SV40 and polyoma. Two months after the close of Biology 211, Berg left the Stanford campus for a sabbatical year with Dulbecco. On his return to Palo Alto in September 1968, his mind had turned away from the prokaryote studies of the past and was concentrated on the effects of viral genes on cell growth and division in eukaryote organisms. He later explained that the purpose of his shift in emphasis was to try to contribute to the prevention, management, and possible cure of the growing number of human diseases that result from defective genes or from errors in translation of the DNA code.

Jackson had joined Berg's lab in 1969, moving over from the Stanford lab of Charles Yanovsky, to whom Jackson had gone from Harvard after a bookish childhood in New York City, where both his parents were librarians. Most of his first year with Berg was spent doing a bacterial experiment that didn't pan out. It was late in 1970, during a discussion of possible successors to the abandoned project, when Berg pointed out the great unexplored area of mammalian genes that might—just might—be manipulated with animal viruses. Had Lobban known about that 1970 talk (no one ever told him) he would have understood why his Ph.D. faculty committee had been unfamiliar with Berg's intent. The simple fact was that the idea didn't emerge from Berg's head until a year after Lobban wrote down the method to carry it out.

Although the Jackson-Symons-Berg report omitted reference to Berg's original intent to grow the SV40-lambda hybrid in *E. coli* in order to use the hybrid as a mammalian cell probe, the cell biology correspondent of the British journal *Nature* recognized that such a culmination would be logical. Writing in one of *Nature*'s November 1972 issues, he said the hybridization of DNA "should be generally applicable with only minor modifications." In other words, "quite unrelated DNAs can be joined." But there was a fly in the ointment which no doubt Berg and his colleagues are well aware of and which may cause them misgivings about proceeding further and could result in their reagent being left permanently in the deep freeze.

> The dilemma is obvious enough: If the SV40 moiety of this hybrid DNA can act as a vector [a vehicle, that is] for the insertion of the

bacterial DNA into chromosomes of animal cells, then no doubt the bacterial and phage moieties can act as vectors for the insertion of the SV40 DNA into *E. coli* chromosomes. . . . But *E. coli* and its relatives happen to constitute part of the human gut flora, SV40 transforms some human cells in culture, and a virus virtually indistinguishable from SV40 has recently been isolated from the brain tissue of people suffering from multifocal leucoencephalopathy [one of the least pleasant deteriorations of the brain and central nervous system].

What would be the consequences if the reagent Berg and his colleagues have made somehow infected and lysogenized *E. coli* in someone's gut as the result of an accident? This possibility, remote though it may seem, can hardly be ignored, and it will be most interesting to learn what criteria the group adopts when it decides whether the scientific information that might be obtained justifies the risk.

Being an efficient craftsman, Stanford Public Information Director Andreopoulos had a standing arrangement with the school's library that brought all journal references to experiments involving Stanford faculty to his prompt attention. On receiving a copy of *Nature*'s commentary on the Berg experiment, Andreopoulos took the copy to Berg and asked whether the time might not be opportune for a statement to the press that would bring the hybrid experiment into the open and possibly avoid a serious misunderstanding later. Andreopoulos had safely negotiated Stanford through the long, hot controversy over Professor Norman Shumway's heart transplant experiments some years before, and was eager to keep the school's nose clean in what he intuitively felt was a potentially explosive situation.

Berg's response was a blast of indignation, directed not at Andreopoulos but at *Nature.* He said the editors of the journal had known that the part of the experiment criticized by the cell biology correspondent had been abandoned, and that *Nature* had denied him due credit for recognizing a problem and acting to eliminate it.

Andreopoulos argued for a clarifying letter to the editor of *Nature.* But Berg, understandably disappointed that his push for early publication had failed of a happier result, declined to entertain the suggestion.

6

PETER LOBBAN'S report on the splicing of DNA molecules of bacteriophage P22 finally appeared in the 1973 issue of the *Journal of Molecular Bacteriology,* months after it was submitted. The excessive delay in its publication was due in part to the *Journal's* press production schedule, which is several times as long as that of the *PNAS.* Professor Kaiser co-signed the paper, which was titled, "Enzymatic End-to-end Joining of DNA Molecules." A very thorough exposition of the techniques and procedures involved in hybridization occupied eighteen printed pages. The total effect, however, was anticlimatic, not only because of the Berg team's prior publication but because the focus of attention had meanwhile shifted to a new discovery.

Some of the EcoRI enzyme that Professor Herbert Boyer had given to Professor Berg was given in turn by Berg to Janet Mertz, the young graduate student whom Berg had sent to Robert Pollack's mammalian cell culture class in June 1971. Berg asked Mertz to study the enzyme's behavior to see if she could notice anything about it that other researchers had missed. With characteristic concentration and thoroughness, she came up with a totally unexpected observation.

What Mertz noticed was that when EcoRI cleaved an SV40 DNA circlet, the free ends of the resulting linear DNA after a short period of separation seemed to be trying to knit themselves back together again. Puzzled by this apparent eccentricity, Mertz asked Stanford's brilliant electron microscopist Dr. Ronald Davis to focus his powerful lenses on whatever was happening.

By dropping the temperature far down to slow the molecular activity, Davis was able to confirm that the cleaved DNA was in fact restoring itself to its normal circularity. As he studied the phenomenon more closely, he could also see that the severed ends of the DNA were not squared off. EcoRI, instead of making cuts in the two DNA strands precisely opposite each other, had cut one of the strands a short distance away from the point of the cut in the other strand. The effect was to give each end of the newly-created piece of DNA a natural one-strand tail. In short, EcoRI had produced "sticky" ends like those that had become familiar in lambda phage DNA.

Mertz and Davis noted further that all the DNA they cleaved with EcoRI, regardless of its origin, came out with "sticky" ends. They concluded that any two DNA molecules exposed to EcoRI could be "recombined" with the help of nothing more than DNA ligase to form hybrid DNA molecules.

Another Stanford researcher, Victor Sgaramella, separately and independently made indentical observations while working with P22, the bacteriophage Peter Lobban had chosen for his Ph.D. study.

These findings were quickly transmitted to Boyer's lab in San Francisco. There, by tagging pieces of DNA with radioisotopes and pulling them through blocks of gel under the influence of electric current and photographing the patterns that appeared when the gel was exposed to ultraviolet light, Dr. Howard Goodman, a Swiss-trained specialist in DNA sequencing, was able to tell the order in which the base pairs of the molecules were strung out in the genetic message. He found that EcoRI identified a particular palindrome, which random distribution places fairly frequently in the DNA instructions, just as certain combinations of letters occur in many English sentences. That palindrome is shown in the sketch below:

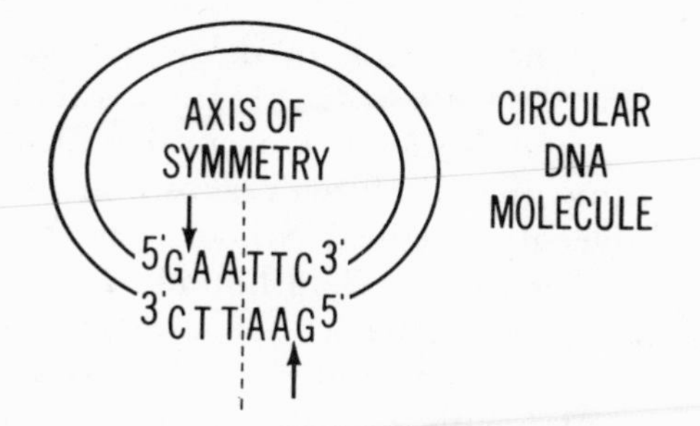

EcoRI did not cleave down the midline of the palindrome, however, but in a mortise-and-tenon zigzag like this:

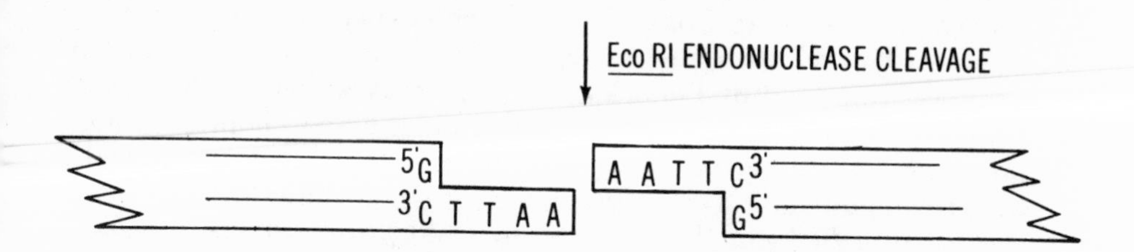

Berg had not known the results of Mertz's mission at the time he submitted the SV40-lambda phage hybrid report to *PNAS*. But when Mertz informed him of it he was so pleased and impressed by her originality and initiative that he omitted his own name in submitting the data to *PNAS* for publication, which occurred late in 1972, in the same issue that carried the announcement of Goodman's confirmation of the base pair sequencing (which was co-signed by Boyer and J. Hedgpeth).

The stage was now set for still greater surprises.

7 PROFESSOR BERG continued to dream of mapping and manipulating the genetic structures that underlie human disease. The widespread fears of SV40 as a probing instrument would delay him appreciably; that much was now plain. But he could not placidly accept the notion of being stopped by ignorance of the unknown. He therefore pushed on with plans for the research biohazards conference he had proposed to Robert Pollack on the telephone.

At Berg's request the initial nomination of delegates was done jointly by Pollack, representing the Cold Spring Harbor Laboratory; Alfred Hellman, representing the National Cancer Institute; and Michael Oxman, representing the Children's Hospital Medical Center of the Harvard University Medical School. All the other conference preliminaries were handled by Berg himself. He used his personal prestige to raise money to transport 100 tumor specialists from all parts of the United States and one from England to the Asilomar Conference Center at Pacific Grove, California, a public park stunningly situated on the Pacific shoreline, and to feed and lodge those 100 persons for three days. He used the administrative apparatus of Stanford's biochemistry department—which for years had been holding a semiannual staff picnic at Pacific Grove—to reserve the Asilomar Conference Center chapel for the conference sessions. He selected the conference delegates from the list of nominees compiled by Pollack, Hellman, and Oxman. He issued the invitations. He welcomed the guests. He made elaborate preparations for taping a transcript of the proceedings, and arranged for the nonprofit publication of the transcript as a limp-backed book, *Biohazards in Biological Research,* by the Cold Spring Harbor Laboratory Press.

Unfortunately, Berg neglected to give the popular press any notice of his intent to hold the conference. Nor did he invite press coverage of any of the sessions, or make available to the media as much as a single sheet of facts about the subject under discussion. Stanford Public Information Director Andreopoulos was discouraged from interesting himself in the affair because "it was not a Stanford event."

Over five years later, the public still knows virtually nothing about that conference, even though the preface to *Biohazards in Biological*

Research strongly suggests a need for popular involvement. The preface says:

> With the widespread interest and growing participation of many laboratories in the problems of animal cell biology and tumor viruses, there is a growing need for consideration of potential health hazards. Much of the experience and knowledge concerning such hazards, imaginary as well as real, is known to only a few people and is not widely publicized. Consequently, there is strong interest on the part of many investigators in holding a series of conferences at which these matters may be reviewed and discussed.
>
> Through the combined sponsorship of the National Science Foundation, the National Cancer Institute, and the American Cancer Society, the first such meeting, the Conference on Biohazards in Cancer Research, was held at the Asilomar Conference Center, Pacific Grove, California, on January 22–24, 1973. This book contains the proceedings of that conference.

"After all, I am something of a ham," Professor Berg once told an interviewer. "I enjoy putting on a show." He was commenting on his performance on the lecture platform, where he is famous for the style with which he projects authority. "I do not enjoy the hard work of organizing the data for a lecture, but the lecturing itself is fun."

In organizing the larger spectacle at Asilomar, he may have been tripped up by impatience with the details of preparation. In any case, he failed to take into account the fact that scientists generally prefer their technical jargon to plain English, while the great mass of nonscientists who constitute society much prefer plain English. Consequently, the stenographers he hired to transcribe the tapes of the Asilomar proceedings were reduced to reliance on phonetic spelling. The result was close to chaos. Pollack, Hellman, and Oxman spent many hours straightening out the mess. Scientists who had delivered prepared papers filled in scores of missing lines, but the ad-lib discussions that followed the papers were not so easy to reconstruct. Berg complained afterward that large chunks of significant observations had been lost.

It seems safe to say that most laymen would not find *Biohazards in Biological Research* worth the $5 charged by the Cold Spring Harbor Laboratory Press for a copy of the book. But there are easily readable passages of importance in the 369-page volume. For example, Dr. Andrew M. Lewis, Jr., of the National Institute for Allergy and Infectious Diseases, made the following remarks about the possible present-day implications of contamination of polio and adenovirus vaccines by SV40 in the 1950s and 1960s:

[Available data] indicate that of the millions of persons who received poliomyelitis vaccines contaminated by . . . [SV40], the records of only 6764 persons have been examined sporadically during the past 15 years. One . . . [study] suggested the need for more careful and continued surveillance. These findings indicate that there has been no systematic, in-depth followup study of persons receiving SV40-contaminated poliomyelitis vaccines and I have been unable to find any followup study of military or civilian personnel who received SV40-contaminated adenovirus vaccines. Thus while there appears to be no untoward short-term effects from human infection with SV40, proper studies evaluating the long-term effects have not been undertaken. As it is now about 15–20 years since the general use of these contaminated vaccines, the time for the appearance of any untoward long-term effects may soon be at hand.

Sharp differences in the value placed on human life in the laboratory were expressed in an exchange of opinions among Dr. Francis Black, of the Yale University School of Medicine; Dr. Robert W. Miller, of the National Cancer Institute; and Dr. James Watson, of the Cold Spring Harbor Laboratory:

Black: The costs of safety precautions will inevitably reduce the number of grants available and increase the time required to reach our ultimate goal.

If we do believe in our mission of trying to control cancer, it behooves us to accept some risk. Even if, as has been suggested, five or ten people were to lose their lives, this might be a small price for the number of lives that could be saved.

Miller: I am alarmed at the thought that some people are willing to accept some deaths in order to accomplish their work. I do not see how you could have predicted in 1950 how many deaths would occur 15–25 years after stilbestrol was given to pregnant women. It was thought then that there was no hazard, and now there are almost a hundred young women who have developed cancer as a result [of stilbestrol's use]. Also, it was suggested that the carcinogenicity of viruses is the only measure of health hazard of mortality. But we have been told here that viruses are related to cerebral degenerative diseases, which are just as fearsome to me as cancer is.

Watson: I'm afraid I can't accept the five to ten deaths as easily as my colleague across the aisle. They could easily involve people in no sense connected with the experimental work, and most certainly not with the recognition and fame which would go to the person or group that shows a given virus to be the cause of a human cancer.

If we wish to broaden the base of tumor virus research by bringing it into academic environments, we will generate a situation where, unless "safe labs" are constructed, many outsiders will be exposed to the many viruses that we want to study at the molecular level. As we all know, "safe labs" will require considerable expense and cannot be funded from research money given to do experiments. They will be built only if specific funds exist for biohazard prevention, say on the order of $125,000 for a moderate-sized lab and much more if we are dealing with a major lab.

Watson's opinion on the erosion of research funds by the costs of biohazard protection was contradicted by Dr. P. G. Stanly, of the National Cancer Institute:

I should like to assure this audience that there is no prohibition on requesting funds in research grant applications for necessary equipment or facilities to reduce biohazards. I know of no grant application that has been turned down merely because such a request was made.

Then there were these illuminating remarks on the nature of risk:

Dr. Wallace Rowe: Much of microbiological safety consists simply of having good habits. For example, it would never cross the mind of a trained microbiologist to touch anything that a drop of virus had fallen on, even if it was the most harmless virus around. Lab workers should have the operating room mentality, that there are clean and dirty areas with clearly defined but constantly changing boundaries.

Dr. Michael Oxman: Curiosity, faith in the biological significance of the expected result, humanitarian impulses, or even personal ambition may convince an investigator to take certain risks. I believe he should be free to do so, as long as only he is at risk. He does not have the right to make that decision for anyone else. This principle of "informed consent" can readily be applied to other investigators, graduate students and technicians, as long as they are informed of the nature and magnitude of the risk and are then free to decide whether or not they are willing to take it. However, the principle of "informed consent" cannot realistically be applied to glassware washers, secretaries, housekeeping personnal, workers in adjacent laboratories or the public at large. Consequently we must insure that under no circumstances will these people be exposed to risks as a result of our research.

Dr. Robert W. Miller: I wonder about the concept of informed consent among people employed to work in virus laboratories. Are they advised that the risk may be different from work elsewhere? Also, do employers advise women workers to notify the employers when the workers are pregnant? These questions suggest the possibility of a formal registry of workers that might contain identifying information on the worker, his progeny, and past medical history. Perhaps new employees could be asked about medication which they take frequently. The responses might draw attention to diseases (such as severe asthma) which might be grounds for exclusion from a position in a virus laboratory.

Epidemiologic studies of virus laboratory personnel and surveillance of human cancer occurences among them were urged by Dr. Philip Cole of the Harvard School of Public Health.

I had hoped that out of the conference would come the formation of a body of responsible persons committed to evaluating the many suggestions that have been made. This group might then formalize and disseminate those suggestions which it can recommend. Irrespective of what epidemiologic studies, if any, are recommended, I would urge that we begin promptly to do the one thing which would enhance any future investigations, epidemiologic or virologic. That is, to establish a population register to include all persons whose work brings them into contact with known or suspected oncogenic viruses (and chemicals) and, possibly, their contacts as well. The information to be registered can be decided after discussion but must include some measures of exposure, health status, and identifying information adequate to allow the individuals to be followed-up.

Professor Berg closed the conference by asking a question that had been repeatedly asked of him: "Where do we go from here? What is the next step after Asilomar?" His answer:

There are several here who would argue that there need be no next step—there is no problem! Admittedly there is no clear-cut [scientifically provable] indication that the oncogenic viruses or cell culture systems currently being studied in laboratories round the world constitute a serious threat to the scientists working with them or to the population at large. That's reassuring. Nevertheless, I'm persuaded by what I've heard that prudence demands caution and some serious effort to define the limits of whatever potential hazards exist. To do less, it seems to me, is to play Russian roulette, not only with our own health, but also with the welfare of those who are less sophisticated in these matters and who depend on our judgment for their safety.

What steps can we take, then? I am persuaded that epidemiologic

> studies may well be the best way we have of detecting a low-order risk; moreover, it has been pointed out that the best chance of detecting such a correlation would be to go to the individuals where the risk is greatest, in the laboratories where people are in contact with these agents most frequently. I would, therefore, like to call for such a study, most logically sponsored by or even carried out within the NIH, to determine if there is increased risk of cancer [or other diseases] stemming from current and projected laboratory researches with biologic oncogenic agents.

In an interview granted some time after the conference, Berg expressed annoyance at the failure of the National Institutes of Health to act on his recommendation. NIH, he said, had excused itself by telling him it was "having great difficulty in devising a questionnaire or intelligence-gathering device that wouldn't upset a lot of people." So strong was its anticipation that "many places probably wouldn't want to provide the information—reporting of accidents, for example—for fear of opening themselves to threat of legal suit" that NIH didn't bother to complete even a feasibility study.

The public was left in ignorance of the fact that it had been deprived of protective measures, which the scientists were prepared to support. Had the Conference on Biohazards of Cancer Research been open to the press, had a press summary even been issued at the end of it, there might have been at least a chance of developing political pressure toward a more responsible attitude within NIH.

8

AROUND THE same time in November 1972 when Professor Berg in the privacy of his laboratory was angrily blasting the editors of *Nature* for perfidy in criticizing his already abandoned plan to grow a hybrid SV40-lambda phage DNA molecule in the *E. coli* bacterium, another Stanford researcher was ravenously attacking a sandwich in a kosher delicatessen within a few steps of Waikiki beach in Honolulu, reputedly the only delicatessen in the Waikiki neighborhood where customers are regularly greeted with the ancient Hebrew welcome, "Shalom!" This famished scientist was Dr. Stanley N. Cohen, a straight-backed, black-bearded, bald-topped, piercing-eyed member of the faculty of Stanford's Medical Department. The hour was late, and he was weary from a long day in a Japanese-American conference then under way at the East-West Center of the University of Hawaii under the sponsorship of the U.S. National Science Foundation. Sharing the delights of the delicatessen with him were three colleagues of long standing: Professor Herbert Boyer, of the University of California (San Francisco) Medical School; Professor Stanley Falkow, of the University of Washington at Seattle; Professor Charles C. Brinton, who was on sabbatical from the University of Pittsburgh; and Brinton's wife, Ginger. Between bites Cohen talked eliptically of a plan he had for creating hybrid DNA molecules without the help of viruses. The method he had worked out in his head would, he said, avoid the risks that had plagued Berg.

Professor Boyer, whose cherubic face, framed by a natural helmet of tightly coiled blond curls, gave no suggestion of his scientific prestige as a bacterial enzyme specialist, was the main object of Cohen's remarks. Cohen was candidly soliciting Boyer's collaboration in an experiment. The others were passive listeners. Except for Ginger Brinton, however, all were expert in extra-chromosomal genetics and understood what Cohen was proposing—that hybrid DNA molecules be constructed from plasmids, which are free-floating circles of DNA that live and reproduce autonomously in bacteria.

After finishing the midnight repast, the five sated but tired friends walked back toward their respective hotel rooms. On the way, Falkow turned to Cohen and Boyer. "Well, fellows," he said, "you can have RSF1010 if it will help you. And if your system works I wish you'd let me know. I'd like to use it in my work with the *E. coli* toxin gene." RSF1010 is a plasmid that conveys resistance to antibiotics on its bacterial hosts (hence the R in its name). The plasmid was discovered by Falkow (therefore the SF). The *E. coli* toxin gene is the gene of the *E. coli* bacterium that causes diarrhea in animals, including man.

For Cohen, the creation of hybrid DNA molecules would be an exotic departure from his accustomed line of research. His main interest had always been in the clinical application of new knowledge. He had studied medicine at the University of Pennsylvania and had done his hospital internship and residency there, later moving to the Albert Einstein Medical School in New York City and then to Stanford in 1968. At Stanford he had been concentrating on exploration, at the molecular level, of the phenomenon of bacterial resistance to antibiotics.

Overprescription of antibiotics by physicians of the United States in the late 1950s and early 1960s was an ill-concealed national scandal, so much so that I, at that time science editor of *Saturday Review,* felt obliged to expose the behavior of the director of the antibiotics division of the U.S. Food and Drug Administration. I had found him accepting pay from drug advertisers for promoting worthless compounds of antibiotics at the same time that the federal government was paying him to protect the American consumer from antibiotic abuse. Under the influence of ecstatic ads published in journals he edited, doctors across the country were prescribing antibiotics for a steadily growing multitude of minor ailments instead of reserving them for treatment of the serious disorders that antibiotics were designed to fight. After a year of pursuit in *Saturday Review*'s pages, I succeeded in forcing removal of the offending official from his post. His departure slowed the antibiotic prescription rate but not until after promiscuous use of those powerful drugs had bred strains of bacteria immune to antibiotic attack.

It was not in the United States, however, but in Japan that the first major thrust toward elucidation of the molecular mechanics of bacterial resistance was undertaken. The action was provoked by a Japanese woman who returned home from a visit to Hong Kong in 1955 with an intractable case of dysentery. After repeated frustrations, her doctor identified the cause of her sickness as a typical dysentery bacillus of the genus *Shigella.* But this particular *Shigella* had extraordinary powers. It successfully resisted four different drugs—sulfanilamide and three antibiotics: streptomycin, chloramphenicol, and tetracycline. And it turned out to be the forerunner of other *Shigellas* that produced a series of severe dysentery epidemics in Japan over a period of years.

In 1959 Japanese researchers demonstrated that the *Shigella* acquired the multiple drug resistance from *E. coli* resident in the human gut, and that the *Shigella* passed the resistance on to other *E. coli* as well as to other *Shigella.* In 1960 the Japanese investigators determined that the exchange did not involve the chromosomes of the bacteria but occurred independently of them. The still mysterious agent was first referred to as an "R factor" and then as an "episome." "Episomes" are by definition "other bodies," vague entities somehow related to bacteria and viruses without being essential to the survival of either.

While working at the U.S. Army's Walter Reed Hospital in Washington, D.C., a scientific team that included Professor Falkow showed that "episomes" were composed of DNA. In 1964 Professor Brinton and some of his colleagues at Pittsburgh confirmed that the "episomes" traveled through bacterial pili, which are thread-thin tubes equivalent in function to the mammalian penis. By 1965 it was clear that the sexual promiscuity of intestinal bacteria was indirectly responsible for the wild growth of antibiotic resistance, which before 1955 had been rarely seen anywhere on earth but a decade later was present in 60 to 70 percent of all intestinal bacteria found in the feces of hospital and medical clinic patients everywhere.

Finally, it became recognized that the "R factor" was a type of plasmid, the extra-chromosomal entity that Joshua Lederberg and one of his students identified in 1952, seven years after Lederberg's discovery that bacteria can have sexual "conjugation."

The plasmids that Lederberg first saw he called F plasmids, the F standing for fertility. The possession of an F plasmid conferred maleness on a bacterium and allowed that bacterium to grow a pilus. If a pilus were sheared off violently in the laboratory, the presence of the F plasmid assured the regeneration of the destroyed appendage.

F plasmids were usually found in the bacterial cytoplasm, outside the chromosome, but sometimes they were integrated into the chromosome, which was a single closed loop about a millimeter long, or a thousand times as long as the bacterium itself. Those F plasmids that were in the cytoplasm were readily transferred to a female bacterium through the pilus, but they made the trip alone. When an F plasmid became integrated into the chromosome, however, the plasmid "mobilized" the chromosome, causing the closed loop to open and pass one of the two strands of its DNA in whole or in part through the pilus to the female. There the inserted DNA strand would be copied to complete an enlarged double-stranded chromosome in the female while the DNA strain remaining in the male would likewise be copied to restore the double-stranded chromosome of the male.

Continuing research showed that although the R plasmid passed through the pili, they did not use the same reproductive apparatus as the F plasmids. F plasmids replicated at the same rate as the bacteria that harbored them. R plasmids replicated much faster than the host bacteria.

By the time Cohen, Boyer, Falkow, and Brinton arrived in Honolulu for the Japanese-American Conference on Bacterial Plasmids in November 1972, bacteria carrying R plasmids were resistant not only to sulfanilamide, streptomycin, chloramphenicol, and tetracycline but also to neomycin, kanamycin, ampicillin, furazolidone, gentamycin, and spectinomycin. In other words, the R plasmids were causing the bacteria to produce enzymes that selectively blocked treatment of dysentery, typhoid fever, cholera, plague, staph, chronic purulent infections, and 90 percent of all urinary infections. The resistant bacteria could be "cured" of their resistance by application of acridine dyes, but the dyes were too toxic for use in clinical medicine. Some other route of attack had to be found, and that route had to follow very fundamental pathways, for powers beyond those of drug resistance were being traced to plasmids. It had been found, for example, that one plasmid releases a poison that kills bacteria that lack that plasmid. Still other plasmids cause tumors in plants, convey resistance to heavy metals and to ultraviolet light, and control manufacture of bacterial enzymes that chop up DNA. In 1972 the rate of discovery of new plasmids was so high that it was impossible to identify them properly without a nomenclature manual. A beginning on such a manual had been made in 1966, and one order of business at the 1972 conference in Hawaii was appointment of an expert panel to finish the job. Cohen and Falkow were nominated to sit on the panel, and both accepted. But their principal purposes for attending the meeting were quite apart from the nomenclature problem.

Cohen's main reason for going to Honolulu was to report the latest development in his part of the search for the key to antibiotic resistance. He had been growing bacteria on laboratory dishes, waiting for them to multiply and then bursting them open with detergent, spinning the plasmids away from the chromosomes in a centrifuge, separating the plasmids from surrounding debris with appropriate chemicals, and finally knocking the plasmids to bits in a blender in order to try to relate the DNA fragments within them to their genetic functions.

A generous supply of duplicate fragments would allow repetitive experiments under identical conditions for comparative purposes. Cohen's provision for such a supply constituted the new development. He and two laboratory assistants—Annie Chang and Leslie Hsu—had discovered that, by modifying a method first tried by Professors Kaiser and Hogness in 1960 and vastly improved by Hawaiian students in 1970, they could treat *E. coli* with calcium chloride and so persuade the bacteria to accept occasional plasmids for replication. The action of the calcium salt wasn't fully understood, but it was believed to inhibit *E. coli*'s restriction enzymes from chewing up the plasmid DNA. In any case, the transformation technique was exceedingly slow. Only one in a million of Cohen's plasmids got through and transferred its DNA. But the resistance to antibiotics was an advantage. For if the host bacteria were grown in the presence of the right antibiotic, only the resistant ones survived.

The efficiency of Cohen's supply system could be greatly improved if all the fragments were identical instead of being randomly sized and shaped as they were when they came out of the blender. How could such uniformity be achieved?

Cohen knew he had the answer as soon as he heard Boyer tell the conference what had been happening in Boyer's lab in San Francisco, an hour's car drive away from Cohen's lab in Palo Alto. Boyer was a Pennsylvania coal miner's son who had moved from the University of Pittsburgh to the medical school of the University of California in 1966, two years before Cohen moved from New York to Stanford. In San Francisco, he began seeking an answer to a question inherent in the transmission of the DNA code from DNA to RNA to the enzymes to the amino acids and the proteins that make up life. That question is: How do the proteins recognize particular segments of DNA? Recognition must precede response to the DNA instructions. "To the rest of the cell, DNA is just a monotonous long skinny thread," Boyer says. "But proteins can tell a particular segment of that long skinny thread from the rest of the thread. How do they do it?"

The answer, Boyer thought, would be found in the enzymes. He got at them by breaking down the bacterial cell wall, which is a very rigid structure, either by bombarding the bacteria with ultrasound waves or grinding them up with aluminum powder. He came out with a paste, which he centrifuged to separate the enzmyes from the cellular debris in which they were embedded.

A host of different kinds of enzymes had to be sorted out and examined. There are billions of them in a cubic millimeter of space. There are at least eighty different restriction enzymes alone. We have already learned that Boyer was not the first to isolate a restriction enzyme. But his graduate student Yoshimori was the first to isolate one that did its cutting in a special pattern that contributed some basic hints about the manner in which proteins identify particular sequences of genes.

By the time of the Honolulu plasmid conference of 1972 both Boyer and Cohen knew of Janet Mertz's discovery (shared independently by Victor Sgaramella) that SV40 circlets severed by Yoshimori's enzyme, EcoRI, had "sticky" ends. They also knew that those findings had been confirmed by Goodman's sequencing of EcoRI's cutting site, as we saw in chapter 6.

As he listened to Cohen's report at the East-West Center, Boyer saw at once that EcoRI would be a much more precise operating tool, and a much faster performer, for use on Cohen's plasmids than the blender at Palo Alto had been. EcoRI had already been supplied to Berg. Why not to Cohen? Boyer made the offer. It did not satisfy Cohen, who had much more than identical DNA fragments on his mind. Cohen thought he saw a clear chance to construct hybrid DNA molecules. If he were to take the chance, he would need the help of Boyer's nimble mind as well as the help of enzymes from Boyer's laboratory. The two men would have to be full and equal partners in the experiment from beginning to end. That was how Cohen saw it. There were some minor hesitations during the subsequent discussions. But the collaboration pact was sealed at the midnight meeting in the delicatessen on Waikiki.

Very soon after their return to California, Boyer visited Cohen's lab to begin the opening phase of the project. The first requirement was an appropriate plasmid. Cohen's earlier plasmid whackings in the blender had produced, among many others, a large, multiple antibiotic resistant plasmid which he had named R (for antibiotic resistant) 6–5. When the DNA from a batch of it was mixed with EcoRI, EcoRI found twelve places where the DNA sequence of R6–5 read $\genfrac{}{}{0pt}{}{\text{GAATTC}}{\text{CTTAAG}}$, and R6–5 was chopped into a dozen pieces. In Cohen's previous knocking about of R6–5, a piece of DNA less than a twelfth as large as R6–5 was isolated. Cohen had named this piece Tc6–5, to indicate that it was resistant only to tetracycline. When left to itself, Tc6–5 curled itself into a circle and became a plasmid in its own right. When Tc6–5 was mixed with EcoRI, EcoRI found only one reading of $\genfrac{}{}{0pt}{}{\text{GAATTC}}{\text{CTTAAG}}$, and the dwarf plasmid was cleaved only once. Because it had retained the genes that enabled it to replicate (as well as the gene that made it resistant to tetracycline), the dwarf was ideal for Cohen's purpose, and he renamed it pSC (for Stanley Cohen) 101.

With that choice out of the way, the two experimenters (it should be mentioned that they were helped by Annie Chang at Stanford and Robert Helling at San Francisco) agreed on a division of labor. The DNA would be removed from the plasmids at Palo Alto and sent to San Francisco, where the DNA of pSC101 and whatever other DNA was chosen as pSC101's cargo would be cleaved by EcoRI and spliced together. The hybrid DNA molecule then would be returned to Palo Alto for insertion into and replication by *E. coli.*

By late spring of 1973, Cohen and Boyer and their aides had completed three splicings of plasmid DNAs, demonstrating progressively greater versatility for the method with each experiment.

At first, only the joining of different pieces of DNA from a single source was attempted. DNA from pSC102, another daughter of R6–5, was joined to the DNA of pSC101. EcoRI cut pSC102 into three pieces. Any one of the three segments would have dovetailed into the cleavage of pSC101. After the joining of one took place, the pSC101–pSC102 hybrid was put back into *E. coli.* Replication followed, and the descendants of the hybrid carried the tetracycline resistance of pSC101 and the kanamycin resistance of pSC102.

The second experiment in the series spliced the DNA of pSC101 to the DNA of RSF1010, the plasmid that Falkow had offered after the midnight snack on Waikiki Beach. Falkow had originally isolated RSF1010 from *Salmonella typhimurium,* a bacillus quite different from *E. coli.* RSF1010 was resistant to streptomycin and sulfonamide. Like pSC101, it had only one $\genfrac{}{}{0pt}{}{\text{GAATTC}}{\text{CTTAAG}}$ sequence in its DNA and therefore was cut only once by EcoRI. The resulting hybrid, which reproduced in *E. coli,* carried RSF1010's resistance to streptomycin and pSC101's resistance to tetracycline.

The third experiment spliced the DNA of plasmid pI258 into the DNA of pSC101. pI258 came from strain 8325 of *Staphylococcus aureus* and was resistant to penicillin, erythromycin, cadmium, and mercury. EcoRI read all the GAATTC/CTTAAG sequences in pI258 DNA and cleaved pI258 into four pieces. Any of the four would have fit neatly into the cleavage of pSC101. The hybrid produced by the joining of one of them to pSC101 replicated in *E. coli* and its progeny carried the powers of resistance of the staph and of pSC101.

In summary, the DNA of pSC101, which had been produced in Cohen's laboratory during experiments with the giant *E. coli* plasmid R6–5, had been joined first to the DNA of another plasmid derived from R6–5 and later to the DNA of plasmids from two organisms very different from *E. coli.* All three organisms were prokaryotes, so it was too early yet to say whether prokaryotes and eukaryotes could be successfully spliced. But the Cohen-Boyer team had gone farther with plasmids than Jackson, Symonds, and Berg had gone with viruses, and some said that whereas the virus method of hybridizing was extremely sophisticated and tedious to duplicate, the plasmid technique was so simple that high-school pupils could easily learn it.

As the senior investigator of the hybrid plasmid project, Cohen assembled the data and wrote drafts of two reports for the approval of Boyer and their helpers. He planned publication in the *Proceedings of the National Academy of Sciences.* The earliest publication date he could hope for, he figured, was November 1973.

9

DR. NEIL E. GORDON walked off the roof of a tall building in the midst of a spell of emotional depression. Events of the moment had persuaded him that he was a failure, and he saw no point in continuing to live. Actually, few other men had influenced the course of modern science as much as he. A chemistry professor at Johns Hopkins University, he was among the earliest to recognize that science had got too big, too hurried, too specialized, too chopped up, and too cooped up to properly encourage speculative thought. In 1931 he gathered about him the best minds he knew in chemistry and its hyphenated offshoots and took them to a secluded island in Chesapeake Bay for a week-long talkfest on scientific frontiers of the future. He was an average-tall, sandy-haired, utterly humorless man, but his stubbornness of purpose had a convincing quality that enabled him to persuade somewhere between twenty and thirty researchers prominent in academic, government, and private industrial laboratories to join him summer after summer for a leisurely week, first in the lounge and later in the boathouse of the fashionable Gibson Island Country

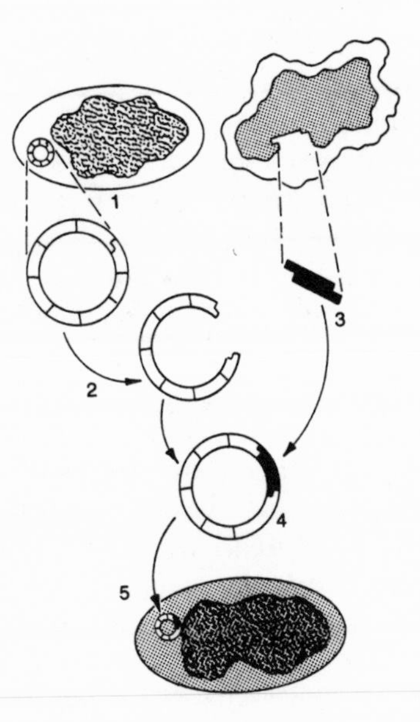

A simplified version of the DNA recombinant technique: 1) A free-floating circlet of DNA, called plasmid, is isolated from a bacterium. 2) The plasmid's ring is broken open. 3) A segment of DNA is taken from the chromosome of another organism. 4) The segment is inserted in the plasmid. 5) The hybrid plasmid is put into a host bacterium capable of reproducing that hybrid. *Chart reprinted from* Business Week *by special permission. :: by McGraw-Hill, Inc.*

Club, surrounded by a thousand acres of wooded hills bounded entirely by water except for a guarded causeway that linked the island to the mainland.

Formal papers were discouraged at Professor Gordon's meetings. Open, free discussion was the purpose. The subject matter was circumscribed to research in progress, incomplete experiments, new theories not yet boiled down and ready for publication. Only two sessions were scheduled each day, one in midmorning, one after dinner at night. The remainder of the time was spent in sporadic discussion on the golf course, on the tennis courts, in the swimming pool, on the bay shore with fishing pole in hand, or out on the water undersail. The trumpet vines, the mocking birds, the osprey's nests, and the mint juleps were just about as far from the conventional conception of a test tube as can readily be imagined. Yet the intellectual ferment was continuous.

Professor Gordon's rule for dress at his meetings was "no jacket, no necktie, shorts if you want to wear them." This shocked the aristocratic members of the Gibson Island Country Club. As long as the conference fees helped to tide the club treasury over financial embarrassment, nothing was said. But as the economy improved the contrast between the regulars and the insurgents began to rankle. To avoid an open clash, Professor Gordon in 1947 shifted his meetings from the muggy Chesapeake to the cool, pond-dotted hills of New Hampshire.

Long before his untimely death, the Gordon research conferences had become a proud ornament of international science, too firmly established to feel more than a momentary shudder at his passing. His faithful administrative aide of many years, Professor Robert Parks of the University of Rhode Island, took his place, and later, when Parks died, he was replaced by Professor Alexander Cruickshank, also of the University of Rhode Island. Not only did the meetings continue, but they spread over the length of the summer, dealing with a different subject every week on each of nine preparatory school and college campuses in New England and California. Attendance continued to be by invitation, and participants in one conference on a given subject would elect the chairpersons for the next conference on that subject.

Although a rule on the point had never been adopted, there was a tacit understanding that all discussions would be strictly scientific. Everything was confidential. No one could be quoted afterward without his explicit permission. No records were kept. Social and political aspects of science were never part of any program—until June 1973, when Dr. Maxine Singer, who happened to have been born in the year when Professor Gordon staged his first conference on Gibson Island, rose from her chair and announced an unscheduled half-hour interval of discussion of a matter of public policy.

It was the morning of Friday, June 15. The last day of sessions of the 1973 Gordon Research Conference on Nucleic Acids was about to begin at the New Hampton School in New Hampshire. Dr. Singer, a research chemist at the National Institute for Allergy and Infectious Diseases at Bethesda, Maryland, since 1968,* was co-chairperson of the conference. She explained that she made her statement with the agreement and at the request of the other chairperson, Professor Deiter Soll of Yale University.

At the time Singer had been elected to the chair in 1971, no one could have foreseen that that event would shape the future history of science. But there were three reasons for things turning out that way. Two of them lay in the nature of primary influences playing on her life. Born into a well-to-do Jewish family in New York City, Maxine

* Now head of the Nucleic Acid Enzymology Section, Laboratory of Biochemistry, National Cancer Institute.

Frank attended Swarthmore College, a suburban Philadelphia school that was run largely on Quaker principles. In an atmosphere saturated with free thought, she there met her husband-to-be, Daniel Singer, and married him in the year of her graduation. While she continued her studies in pursuit of a Ph.D. in biochemistry at Yale, he became a lawyer with public interest leanings and a special concern for the ethics of biology that carried him into association with the Institute of Society, Ethics and the Life Sciences, created with the support of several of the leading philanthropic foundations of the country and headquartered at Hastings-on-the-Hudson in New York State. His activities in the public arena meshed with those of his wife after she left Yale to accept a U.S. Public Health Service post-doctoral fellowship with the National Institutes of Health. There at Bethesda, she was caught up in the various causes of the Federation of American Scientists, an organization of thousands of natural and social scientists and engineers that had first been formed in 1946 as the Federation of Atomic Scientists to function as the conscience of the scientific community. As such it was licensed to lobby on all matters related to the problems of science and society. Among other ventures, she had stuffed envelopes in the FAS drive against the nomination of arch-conservative former Atomic Energy Commissioner Lewis Strauss for the post of U.S. Secretary of Commerce. Through her involvement in that near-disastrous affair, Daniel Singer had become legal counsel for FAS.

At the age of forty-two, Dr. Maxine Singer was the mother of four children, a warm and vivacious, highly intelligent, open-minded rarity in the scientific community, flexible enough to respond creatively to the unconventional pressures that were the third reason why her co-chairing of the 1973 Gordon Research Conference on Nucleic Acids would make history. Those pressures arose spontaneously from the scientific experiment that Professor Herbert Boyer, of the University of California (San Francisco) Medical School, had reported to the conference on the previous day.

Stanford Professor Stanley Cohen, senior collaborator in the work that Boyer described, would have preferred not to release any information about the plasmid-generated DNA hybrid molecules that they had constructed until after publication of their laboratory procedures in *Proceedings of the National Academy of Sciences.* That publication was not scheduled until November 1973. Knowing of Boyer's invitation to attend the Gordon Research Conference on Nucleic Acids in June, Cohen discussed the matter with Boyer and obtained agreement that nothing would be said at the New Hampshire meeting. Gordon Conference enthusiasm for anything new and promising is infective, however, and when Boyer's turn to speak came on next to the last morning of the conference, at a session chaired by Professor Daniel Nathans and titled "Bacterial Restriction Enzymes and the Analysis of DNA," the ebullient young scientist from San Francisco got up and happily spilled the beans.

Perhaps he was subconsciously subdued by his unpremeditated departure from the agreement he had made with Cohen. In any case, Boyer did not convey the true significance of the plasmid experiment to most of his audience. No discussion whatever followed his report. Not until a later speaker at that morning's session, Dr. William Sugden, a thoughtful Cold Spring Harbor Laboratory graduate who helped to design a method of "fingerprinting" genes and had been a spasmodic attendee at Robert Pollack's tumor virus workshop class in 1971, said of Boyer's data, "Well, now we can put together any DNA we want to" did the meeting hall begin to buzz with excitement. Singer herself later confessed that she missed the main point of Boyer's presentation until other conferees sought her reaction to Sugden's remark.

One aspect of what followed is puzzling. Sugden had said nothing new. An earlier chapter of this book quoted an almost identical statement by the cell biology correspondent of *Nature* in a report on the Berg team's work in November 1972. "Quite unrelated DNAs can be joined" was *Nature*'s published judgment then.

Why did the splicing of plasmids arouse such a knowledgeable audience as a Gordon Research Conference when there had been so little reaction to the hybridizing of the SV40 virus and the lambda phage? Could it have been that the message needed to be repeated before its full meaning sank in?

One clue seems to corroborate a positive answer. Edward Ziff had been among the scientists who discouraged Berg from completing the SV40-lambda phage experiment. And it was Ziff and Paul Sedat, who, like Ziff, was then a visiting American scholar at the Medical Research Council Laboratory of Molecular Biology in Cambridge, England, approached Singer just before the start of the evening session of the Gordon conference that followed Boyer's talk. "Shouldn't there be some discussion of potential dangers inherent in this new technology?" they asked.

Given the traditional Gordon conference rules on confidentiality, there is no ethical way to search for the reasons these two young men gave to Singer for their concern about DNA hybrids. But it is possible to read the questions Ziff posed in reporting on the Gordon conference action in the *New Scientist* of London on October 25, 1973.

> Such [DNA hybrid] molecules constitute new combinations of genetic material. Thus their biological properties remain untested, either in the

> laboratory or in nature. Many sorts of DNA hybrids are likely to prove entirely innocuous. However, certain cases have aroused concern. Some viral DNAs are capable of transforming cells in tissue culture to a malignant state. . . .
>
> So far no danger has been established. . . . It remains possible that an animal virus DNA cannot be maintained in a bacterium. And humans may have sufficient mechanisms for protecting themselves against these foreign DNAs. Until these possibilities are tested, however, a conscientious scientific community will give especially careful attention to the means of its research as well as its goals, and take the full precautions necessary for ensuring public safety.

Of particular interest to the research community just then was the large-scale production and isolation of viruses that cause cancer, Ziff's *New Scientist* article pointed out. Those viruses "often must be . . . purified in concentrated form. The handling of concentrated viruses can present special problems of dissemination, such as the formation of aerosols [micro-droplets of airborne virus]. No virus has yet been conclusively demonstrated to cause concern in man. The effects of exposure to a carcinogenic factor, however, may not be evident until years afterwards. . . . Therefore . . . new safety guidelines will be required, and adequate protective facilities provided. . . . Until such time as reasonable guidelines are established, vigorous discussion and debate within the scientific community will continue to provide the greatest safeguard of the public good."

However differently Ziff may have stated his case to Singer at the Gordon conference in June, she heard his argument with Professor Soll at her side. Professor Nathans and a scientific colleague with an unusual set of qualifications, Dr. Sherman Weissman of Yale, were within earshot. Nathans and Weissman both wondered aloud whether an open debate might not lead to undesirable restriction of the traditional freedom of scientific research. After urging them to consider the matter further, Singer went into a private huddle with Soll. A brief exchange was enough to persuade the two of them that if level-headed young people like Ziff and Sedat could react as strongly as they had reacted, the wise course would be to open the concluding conference session next morning with a half hour of what would amount to political debate before completing the previously set scientific program.

At Soll's request, Singer drafted a brief statement that night. Many later commented that if Nobel laureate Joshua Lederberg had got the assignment, the question the youngsters raised would almost certainly not have reached the floor. But Singer moved in the opposite direction and, in effect, challenged her colleagues to stand up and be counted on the much mouthed question of social responsibility. Her opening remarks are preserved in a letter she wrote as a matter of record later:

> First I will describe briefly the question that has been raised by some participants in the conference, as I see it. We all share the excitement and enthusiasm of yesterday morning's speaker who pointed out that the scientific developments reported then would permit interesting experiments involving the linking together of a variety of DNA molecules. The cause of the excitement and enthusiasm is twofold. First, there is our fascination with an evolving understanding of these amazing molecules and their biological action, and second, there is the idea that such manipulations may lead to useful tools for alleviation of human health problems. Nevertheless, we are all aware that such experiments raise moral and ethical issues because of the potential hazards such molecules may engender. In fact, potential hazards exist in some of the viruses many of us are already studying. Other problems will arise with hybrid molecules we are contemplating. Furthermore, these hazards present problems to ourselves during our work and are potentially hazardous to the public.
>
> Because we are doing these experiments, and because we recognize the potential difficulties, we have a responsibility to concern ourselves with the safety of our coworkers and laboratory personnel as well as with the safety of the public. We are asked this morning to consider this responsibility.
>
> I fully understand that I have not discussed this topic exhaustively, and that there are even arguments to be made about the factual content of my statement. However, having the problem raised so late in the meeting requires that we deal expeditiously with this question. As Chairman, I will not permit substantive discussion of the problem but only proposals concerning possible action or inaction on the issue. In fifteen minutes the discussion will be closed and we will vote by a show of hands on any proposals that have been made and seconded. We will then proceed with the full scientific program planned for this morning.

Three proposals were moved and seconded. The first suggested that the conference address a letter of concern to the National Academy of Sciences and to its affiliate, the Institute of Medicine. The second suggested that a letter be drafted and signed by as many individuals as wished to do so. The third suggested that as many individuals as wished to do so write individual letters to the academy and to the institute.

Two ballots were taken. On the first, 78 of the 95 persons present on that Friday morning approved the dispatch of a letter from the conference to the academy and the institute. On the second ballot, 48 persons voted in favor of and 42 voted against sending a copy of the letter to *Science,* journal of the American Association for the Advancement of Science.

Forty-seven of the 142 scientists who had originally enrolled for the conference had gone home before the vote was taken. That was more than a third of the total. To avoid any possible misunderstanding about the legitimacy of the balloting, Singer on June 21 sent a letter to every person enrolled in the conference. "Those who attended Friday morning's session will know what [this] . . . is all about," the letter began. "For those who were not present on Friday morning, the following remarks, which I made then, will explain the issue." Then, after repeating the three-paragraph statement she had made as a preface to the vote, and after giving the outcome of the voting, Singer concluded:

> Because many participants had left by Friday, it was decided to send out a draft of the letter to everyone. It is enclosed. Please send any suggestions for revisions to me by July 15. Also, please indicate below your approval or disapproval and mail to me by the same date. I will assume that anyone remaining silent agrees with the majority and has no serious objection to the draft of the letter.
>
> Of course, these problems are not peculiar to the United States. Those of you who are abroad may wish to send our letter, or a similar one, to appropriate organizations in your own country.

Under a dotted line beneath Singer's signature were boxes labeled "Approve" and "Disapprove" opposite these two statements: "(1) Send letter to Academy Presidents" (sic) and "(2) Publicize letter in SCIENCE Magazine."

Sixty-one comments came back, including an unsolicited one from Paul Berg. Forty okayed publicity for the letter, half as many took an opposite stance, and one couldn't decide one way or the other. Accepting Singer's stated assumption that silence meant consent, all but twenty of the 142 original enrollees at the 1973 Gordon Research Conference on Nucleic Acids favored writing the letter from the conference to the president of the National Academy of Sciences and the Institute of Medicine and asking *Science* to publish the letter. In consultation with Soll, Singer revised the letter to encompass the suggestions of the correspondents. Then she and Soll, as co-chairpersons of the conference, signed the finished document and sent copies to its three destinations.

Because private meetings of individuals and groups are such an established feature of Gordon research conferences, the other 1973 nucleic acid conferees had no occasion to notice a brief encounter between Boyer and one of Berg's graduate students, John Morrow. Morrow took the initiative in order to compliment Boyer on the report of the work with Cohen. After acknowledging the pleasantry, Boyer said, "What we need now are some eukaryote cells." Thoroughly sophisticated in molecular biology, Morrow knew at once what Boyer was getting at. All the hybrids that Cohen and Boyer had constructed were exclusively prokaryote. The big question now was whether the prokaryote *E. coli* would accept and replicate prokaryote and eukaryote hybrids. "I have just what you want," Morrow said. "As you know, I'll be leaving Berg's lab soon to work with Don Brown at Carnegie. Brown intends our first experiments to be with *Xenopus laevis.*" *Xenopus laevis* was an African toad famed as a genetic freak; it had been intensively studied for years. "It will be no problem for me to get you some *Xenopus laevis* DNA." Boyer smiled his cherubic smile and went home to California to give Cohen the good news.

10

WHEN DR. PHILIP ABELSON, the editor of *Science,* got his copy of the letter from the 1973 Gordon Research Conference on Nucleic Acids, he telephoned Singer, who was a member of the editorial board of *Science,* and asked brusquely, "Do you really want to do this?" Singer replied: "A majority of the participants in the 1973 Gordon Research Conference on Nucleic Acids wants to do it." Although it has been rumored that Abelson opposed publication of the letter, Singer denies he did anything more than voice his opinion that publication would bring difficult problems. Whatever Abelson's intent may have been, persons familiar with the production schedule of the magazine were not impressed by the speed with which the letter was prepared for printing. It did not reach the letters pages of *Science* until the issue dated September 21, 1973. Headed "Guidelines for DNA Molecules," the letter was prefaced by a paragraph explaining that "those in attendance at the 1973 Gordon Research Conference on Nucleic Acids voted to send the following letter to Philip Handler, president of the National Academy of Sciences, and to John R. Hogness, president of the National [sic] Institute of Medicine. A majority also desired to publicize the letter more widely." The text of the letter read:

> We are writing to you, on behalf of a number of scientists, to communicate a matter of deep concern. Several of the scientific reports

presented at this year's Gordon Research Conference on Nucleic Acids (June 11–15, 1973, New Hampton, New Hampshire) indicated that we presently have the technical ability to join together, covalently, DNA molecules from diverse sources. Scientific developments over the past two years make it both reasonable and convenient to generate overlapping sequence homologies at the termini of different DNA molecules. The sequence homologies can then be used to combine the molecules by Watson-Crick hydrogen bonding. Application of existing methods permits subsequent covalent linkage of such molecules. This technique could be used, for example, to combine DNA from animal viruses with bacterial DNA, or DNA's of different viral origin might be so joined. In this way new kinds of hybrid plasmids or viruses, with biological activity of unpredictable nature, may eventually be created. These experiments offer exciting and interesting potential both for advancing knowledge of fundamental biological-processes and for alleviation of human health problems.

Certain such hybrid molecules may prove hazardous to laboratory workers and to the public. Although no hazard has yet been established, prudence suggests that the potential hazard be seriously considered.

A majority of those attending the Conference voted to communicate their concern in this matter to you and to the President of the Institute of Medicine (to whom this letter is also being sent). The conferees suggested that the Academies establish a study committee to consider this problem and to recommend specific actions or guidelines, should that seem appropriate. Related problems such as the risks involved in current large-scale preparation of animal viruses might also be considered.

Maxine Singer
Room 9N–119, Building 10
National Institutes of Health
Bethesda, Maryland 20014

Dieter Soll
Department of Molecular Biophysics
and Biochemistry
Yale University
New Haven, Connecticut 06520

Singer later lamented that a key word was dropped from the bottom of the letter. The word was "co-chairpersons," which would have identified the signers of the document as spokespersons for a Gordon conference rather than simply as two interested individuals.

Other readers of *Science* were not so concerned about that as they were disappointed by the technical cast of the letter. For example, how many nonscientists would know what it means "to generate overlapping sequence homologies at the termini of different DNA molecules" and then use those "sequence homologies . . . to combine the molecules by Watson-Crick hydrogen bonding?" As one science writer wryly put it, "The scientists were talking to themselves, over the heads of readers unversed in science, as parents do when they discuss adult subjects in the presence of children."

11

HAD DR. DAVID HAMBURG, a psychiatrist on leave from the faculty of Stanford University, been president of the Institute of Medicine in 1973, as he is today, the remainder of this book might be written in a quite different vein. For Dr. Hamburg is very much alive to science's need for public understanding and trust. And he has publicly stated more than once that the time has passed when scientists can expect to unilaterally dictate decisions on issues of public policy that arise from scientific discoveries and their technological consequences. Judging from the attitudes he has expressed on other occasions of similar import, he almost certainly would have taken some constructive action on the letter from the Gordon Research Conference on Nucleic Acids.

But Dr. Hamburg was not president of the Institute of Medicine in 1973. Dr. John Hogness was. When he received the institute's copy of the letter signed by Singer and Soll, he turned it over to the National Academy of Sciences, deferring to the academy's parental relationship to the institute.

12

LEGEND SAYS that the National Academy of Sciences was invented by Abraham Lincoln to help steer the United States of America safely through a potentially disastrous crisis of identity. According to this piece of folklore, Mr. Lincoln was aided and abetted by farseeing congressmen who understood the role of science in human enlightenment. That it didn't happen that way; that the academy actually was conceived by a scientific cabal and authorized by a law passed unwittingly in the dying days of a lame-duck Congress; that the one legislative sponsor of the law never showed any interest in science before or after; that the first reaction of scientists outside the cabal was outrage over the betrayal of democratic principles in provision of a self-perpetuating academy membership—these historical details are overshadowed by the practical fact that Mr. Lincoln did sign the law and so became the academy's creator.

Throughout the eleven decades that passed between that event and the arrival of the letter from the Gordon Research Conference on Nucleic Acids, the academy had vacillated between two philosophical extremes, one extreme being a belief that society is obligated to science and the opposite extreme a belief that science is obligated to society. In the early summer of 1973, the sense of obligation to society was in the ascendancy, in part because of the initiative of consumer advocate Ralph Nader, who was nudging science writer Philip Boffey to get on with the writing of a book (published two years later under the title *The Brain Bank of America)* that would document the situation for scientists and nonscientists alike.*

A new academy president, Duke University biochemistry professor Philip Handler, had taken office in 1969 with a commitment to restructure the institution to better serve the people's needs. Most significantly, he had created a small, entirely new enclave within the academy labyrinth, named it the Academy Forum, made it directly responsible to himself, chosen as its director Dr. Robert White, a self-proclaimed maverick who exulted in refusal to conform to any "academy type," and instructed White to see that the forum served the academy as an effective vestibule to the outer world.

Much time and effort went into making this vestibule sturdy enough to withstand the political vandalism that occasionally sweeps the academy. Handler recognized that ritualists on the premises would resent any passageway that might serve as an entrance for the common people, and would rarely if ever use it for fear of calling attention to it. So, to chair an Advisory Committee of academicians empowered to generate and maintain forum policies, the new academy head chose a man with cross-cultural credentials, anthropologist Robert McC. Adams, head of the social science division of the University of Chicago.

Professor Adams could not have been more apt for the job that needed to be done. Among other things, he was chairman of the board of the *Bulletin of Atomic Scientists,* which existed for the sole purpose of reminding everyone of the lamentable consequences that followed failure of the desperate last-minute attempt by University of Chicago nuclear scientists to prevent the destruction of Hiroshima and Nagasaki by atomic bombs in 1945. The editors of the *Bulletin* were dedicated to the proposition that the next time such a historic

* Nader picked Boffey from the news reporting staff of *Science.* Boffey now sits on the editorial board of the *New York Times.*

crisis arose, recurrence of disaster might be averted if the scientists intervened aggressively at a sufficiently early stage of events. And now, in 1973, the next time had come, implicitly proclaimed in the text of the letter from the Gordon Research Conference on Nucleic Acids.

Perhaps academy president Handler wasn't as deeply committed to communication with the lay public as he liked to think. Perhaps he experienced a failure of nerve. For whatever cause, he bypassed the Academy Forum (which had already successfully staged one broad public debate on a related subject—the nature of risk) and thereby forfeited a seldom available opportunity to win the confidence of all segments of American society in one dramatic sweep; instead, he referred the Gordon Conference letter to another organizational invention of his, the academy's new Assembly of Life Sciences.

The letter was discussed at the very first meeting of the assembly in September 1973, Dr. Artemis Simopoulos, executive secretary of its division of medical sciences, recalls. Dr. Paul Marks, chairman of that division, led discussion of the issue: Was this a general question of laboratory quality control or was it a specific question that dealt strictly and solely with issues that could emanate from uncontrolled research on recombinant DNA? An expert opinion was wanted. Dr. Maxine Singer, a specially invited guest, was asked to give her views. She suggested that Professor Berg be consulted because of the breadth of vision implied by his earlier renunciation of the final step in his SV40-lambda phage-*E. coli* experiment.

According to Berg, his first word from the academy came in a telephone call from Leonard Laster, an Assembly of Life Sciences staff functionary who reported Singer's recommendation and asked Berg's reaction. Still feeling bruised by the disappointing outcome of the January conference at Asilomar ("It showed us how little was known"), Berg replied that he didn't want to talk off the top of his head and, more importantly, wasn't willing to carry the responsibility for any course of action on his shoulders alone. But he said he would be willing to poll some of his colleagues.

That decision, Laster indicated, would be acceptable to the assembly, and Berg flew east to lunch with Nobelist James Watson and Harvard Professor Emeritus John Edsall, a universally admired elder statesman of American science. After those conversations, Berg called Laster and told him of Berg's decision to form a committee of eight to ten scientists, known to and trusted by him, to pursue the dilemma posed by the Gordon Research Conference letter. Laster assured Berg that the Assembly of Life Sciences would guarantee the committee's expenses, but cautioned Berg not to refer to the committee as a committee of the academy. The academy followed a rigorous method of checks and balances in selecting its committees, Laster explained, whereas Berg was proposing simply to empanel some of his friends.

For the first time in his professional career, Berg felt that he had been disowned. He wasn't sure how to react to the unique situation.

13

DR. RICHARD ROBLIN was an assistant professor of microbiology and molecular genetics at Harvard Medical School. Son of a biochemist and designer of prescription drugs, he had been brought up on the principle that science does not exist for the sake of science but for the purpose of serving the needs of mankind. He had been drawn to molecular genetics because of its potential as a social tool, and he published occasional commentaries on the subject in professional journals. These writings brought him an invitation from the Biophysical Society Ethics Committee to participate in a Symposium on Genetic Manipulation at Minneapolis on June 4, 1974. Wishing to be up to date as well as accurate in his remarks when the time came, he telephoned Dr. Maxine Singer in Washington early in February 1974 to learn what response there had been to the letter addressed to *Science* by attendees at the 1973 Gordon Research Conference on Nucleic Acids. Singer told him that in the previous October the executive committee of the National Academy of Sciences' new Assembly of Life Sciences had voted to set up a committee to look into the concerns expressed by the Gordon letter. She said she thought that Professor Paul Berg of the Stanford University Medical School might be working with this committee, and suggested that Roblin get in touch with Berg.

Roblin had met Berg in the late 1960s in Renato Dulbecco's laboratory at the Salk Institute, where Berg was then a visiting scholar. As he and Berg had got along well together, Roblin gladly followed Singer's suggestion. In writing to Berg, he asked for all available information suitable for educating biophysicists and biochemists in the newest developments in DNA splicing. Berg replied promptly, saying he had had some conversations with a Leonard Laster at the academy but nothing definite had come of them. Roblin therefore wrote to Laster, putting the direct question to him: "Can you tell me whether the [academy] committee has been formally constituted yet, which questions it plans to consider, and who is on the committee?"

The letter to Laster crossed in the mails another letter from Berg to

Roblin. This time Berg said he would head a committee of "eight or ten" to discuss "whether the academy has a role [in supervision of gene splicing] and what it is." That communication was followed by a third in which Berg proposed an alternate stance for the committee, the agenda to call for consideration of "whether or not there is a serious problem growing out of present and projected experiments involving the construction of hybrid DNA molecules in vitro. If a problem exists, then what can be done about it: both short- and long-term actions?" Topics for discussion under that agenda would include hybrid methodologies, experiments in progress or pending, the need for future regulation and, if need were recognized, how to fill it.

Berg asked for Roblin's comments and invited him to attend the organization meeting of the committee to get information for the Minneapolis symposium at firsthand. There were three possible meeting sites—MIT, Rockefeller University in New York City, or Cold Spring Harbor Laboratory on Long Island—and two possible dates, April 16 and 17. Would Roblin express his preference? Roblin opted for MIT on April 16, and offered two suggestions. The first was to forget about debating the academy's role—a decision had already been made on that. Go with the second agenda, Roblin advised, and appoint to the committee "someone with a more radical or skeptical view of the possibilities [of gene splicing]." The "someone" Roblin had in mind would be "like Leon Kass or Jon Beckwith."

Leon Kass, who then was in Washington and is now on the faculty of the University of Chicago, has long been respected for the breadth and depth of his understanding of the interrelationships between science and society. He helped to bring about the creation at Hastings-on-the-Hudson of the Institute of Society, Ethics and the Life Sciences, with which Roblin then was and still is associated. Kass acted as a strong catalyst on Roblin's thought. A question that frequently hung in Roblin's mind after their conversations was: "Can recent advances in biomedical research continue to be viewed as an unalloyed benefit?" Jonathan Beckwith (to restore his full first name) had answered that question in the negative. And Beckwith had acquired major stature as a prophet in Cambridge and elsewhere because the seemingly extreme fears of biologically imposed behavior control that he had voiced at the 1966 press conference celebrating isolation of an entire natural gene for the first time had been so smashingly justified by the subsequent disclosures of mind-controlling drug experiments by the CIA.

Roblin, whose face leaves an impression of even greater asceticism than Beckwith possesses, was active in the Science for the People movement organized in the MIT-Harvard complex by Beckwith and others. Indeed, Roblin participated in the XYY controversy that stopped (through a court injunction) experiments intended to demonstrate the truth or falsity of a postulation that bearers of XYY genes (that is, those who possess an extra Y) end up in institutions for the criminally insane ten to twenty times more often than other people do. Roblin did not go all the way with Science for the People philosophy, but he did see the danger of too tightly squeezing citizens into an identical mold rather than adjusting social and political norms to reflect the variety of a democratic population. At the very least, Roblin felt, no harm and possibly much good might result from incorporating a broad range of views into the Berg committee.

Berg's reaction to Roblin's suggestion was indirect but unmistakable. Neither Kass nor Beckwith nor anyone like either of them was among the group that appeared at MIT on April 17 in response to the invitations Berg issued. Those who did come had remarkably similar academic and professional backgrounds: David Baltimore, of MIT; Herman Lewis, of the National Science Foundation; Daniel Nathans, of Johns Hopkins University; James D. Watson, of the Cold Spring Harbor Laboratory; Sherman Weissman, of Yale; Norton Zinder, of Rockefeller University, and chairman Berg. The one whose experience brought him closest to fulfilling the purpose that Roblin unsuccessfully espoused was Weissman, who was a physician to begin with and deeply engrossed in the social scene through his clinical research on sickle cells, a special hazard to black people. From the molecular perspective, Weissman's work in replicating part of the human globin gene had carried him farther than any of the others into the massive problem of eukaryotic cell behavior analysis.

When Roblin arrived from his laboratory at the Massachusetts General Hospital, he discovered that he was not there in the role of disinterested observer, as he had expected to be. He was a full fledged member of the committee. More than that, he was the committee secretary. Four pages of his tiny script constitute the official record of the so-called "Berg committee's" first day of deliberations, which began at 9:00 A.M. in a room in the library of the MIT Center of Cancer Research, ran through lunch at the faculty club, and continued until 5:00 P.M..

Within the first hour, a favorable vote was taken on Watson's proposal that an international conference be held to obtain an informed global consensus on "what is doable" in DNA molecule splicing. Because of the time that would necessarily be occupied by the preliminaries, the earliest practical date for such a meeting would be February 1975. Because of Stanford's established ties with Pacific Grove, the beautiful California state seaside park, the most convenient place would be the Asilomar Conference Center.

Berg brought with him an important piece of news that hadn't yet got outside the Stanford campus. It suggested that a February 1975 conference on DNA splicing might come too late to provide any leverage toward quieting the doubts expressed by the 1973 Gordon Research Conference on Nucleic Acids. The news was that Berg's student, John Morrow, working with Boyer and Cohen, had replicated functioning *Xenopus laevis* genes in *E. coli.* Apparently the plasmid technique of molecular hybridizing had a broader range of effectiveness than the original Cohen-Boyer experiments indicated. "If we had any guts at all," Berg later quoted Zinder as saying, "we'd tell people not to do these experiments until we can see where we are going." Roblin's notes show Berg himself proposing that until the Asilomar conference could do a definitive job of sorting out relative hazards, an immediate public statement of caution should be issued for the purpose of slowing down or if possible stopping a few types of experiments about which the committee had serious concern. Watson asked if the *Xenopus laevis* genes had been expressed in *E. coli.* In laymen's language, this meant: Had the toad genes replicated by *E. coli* produced the proteins that toad genes would normally produce in the toad? When Berg said no, Watson suggested that it might be premature for scientists to be too deeply concerned until foreign genes grown in bacteria did express themselves. However, the others on the committee agreed with Berg that there ought to be an interim public warning to cover the fact that pSC101 soon would almost certainly have competition from other plasmids which would circulate outside any control system that Cohen might be able to set up at Stanford. Roblin was left with the task of drafting an open letter that would:

1. Announce the Asilomar conference;
2. Urge the director of the National Cancer Institute to undertake production and testing of naked DNA molecules under stringent isolation conditions to determine the danger of infection from them;
3. Pledge the members of the committee, as individual signers of the letter, not to participate in the making of hybrid molecules that might spread cancer or widen bacterial resistance to antibiotics.

The last point was phrased to accommodate Leonard Laster's instruction to Berg not to designate the Berg committee as a committee of the National Academy of Sciences. Some resentment over Laster's words had been expressed during the committee meeting, and there was even a discussion of the desirability of attaching the committee to the American Society for Microbiology. In the end, however, it was agreed that academy prestige was such that even a tenuous relationship with it had great value in the public mind.

14

THE GORDON Research Conference letter put Professor Stanley Cohen in an awkward spot. He had striven to so manage events that the news of his collaboration with Professor Boyer in construction of hybrid DNA molecules from plasmids would not break into the open until the formal description of the work appeared in the November 1973 issue of the *Proceedings of the National Academy of Sciences.* A certain prestige attaches to public notice deriving from *PNAS* publication, and Cohen looked forward to that. Rightly or wrongly, he felt that the Biochemistry Department of the Stanford Medical School looked down its nose at the Medical Department, to which Cohen belonged. This time there would be no excuse for condescension.

Instead of being written in satisfyingly large headlines, however, word of the plasmid gene-splicing achievement had been dribbling along the grapevine unseen ever since Boyer's confidential report to the Gordon meeting. Molecular biologists from everywhere began telephoning and writing supplications to Cohen weeks before the letter drafted at the New Hampshire conference got into print. Requests for seed of the pSC101 plasmid were cluttering his laboratory office desk top. He had resolutely said "no" to them all—including one that had come personally from Professor Berg. Before he would be willing to turn his star performer loose, Cohen wanted to see more clearly how the academy would respond to the suggestion that guidelines be drafted to regulate experiments with hybrid DNA molecules. Although there was no more legitimate cause for it than there had been for Watson's blast at young Dr. Andrew Lewis at the Cold Spring Harbor Laboratory in August 1971, resentment against Cohen's attitude was showing itself. Cohen was not enjoying the experience.

Looking for some form of relief that could be obtained respectably, he telephoned Dr. Maxine Singer and asked her what, if anything, the academy had done about the letter he had read in *Science.* Singer told him the academy's Assembly of Life Sciences had asked Berg for advice. She suggested that Cohen call Berg, and Cohen did so. Might there, he asked, be a place on the committee for him, inasmuch as it was his work with Boyer that had stirred up the excitement? Berg said no; the committee would be made up exclusively of cancer specialists.

Clearly having no claim for designation in that category, and foreseeing no guidelines for some time to come, Cohen capitulated to the pressures surrounding him. About six weeks before publication of the *PNAS* report, he started giving the pSC101 plasmid to any qualified scientists who sought it—provided only that the recipient sign a guarantee that he would not use the plasmid to introduce tumor viruses into bacteria nor to create antibiotic resistant bacterial combinations not known to exist in nature. A further proviso said that experimenters who got the pSC101 from Cohen would not give any of its progeny to anyone else.

In May 1974, Cohen was obliged to visit Cambridge, Massachusetts. As a member of a government science advisory panel, he had to pass judgment on the worthiness of a project funding application. Scientists at MIT who knew of this commitment asked if he would take that opportunity to give an MIT seminar on the use of plasmids in splicing DNA molecules. Cohen accepted the invitation for April 18 and arrived at MIT on that day entirely innocent of the fact that the Berg committee had met at MIT the day before.

A member of the committee, David Baltimore, attended the seminar and afterward talked to Cohen about the committee meeting. Cohen said Berg had told him that all the committee members were students of cancer. Baltimore confirmed that to be the truth, but in relating the committee deliberations he mentioned that one of the subjects of discussion was plasmids. Cohen wanted to know how there could be intelligent discussion of plasmids when the committee had no member who knew anything about plasmids. Baltimore said the committee intended to seek advice from a plasmid nomenclature panel that had been appointed at a Japanese-American conference on plasmids in November 1972. Richard Novick, chief of the department of plasmid biology of the Public Health Research Institute of the City of New York, was chairman of the panel. Cohen was puzzled. He was a member of that panel, but had heard nothing of its intention to advise the Berg committee.

Cohen's confusion grew as he mulled the matter on the homeward plane. Berg had resented Cohen's unwillingness to give up pSC101 when Berg asked for it. There had been some angry words between the two men. Cohen could not help wondering whether his determination to maintain professional independence had anything to do with the contradictory information he was getting about the Berg committee. He couldn't reach any clear conclusion on that question, but it seemed very clear to him that an uninformed committee with the very best of intentions could unwittingly penalize the line of plasmid work on which Cohen had embarked. He decided he could not afford to risk such a happening. He would write a letter of his own to protect himself against any possible adverse effects of the Berg committee's decisions. Upon reaching home, he called Boyer to alert him and then began drafting the letter.

Cohen's office at Stanford is just two floors below Berg's office. Word of Cohen's unusual activity went quickly upstairs. Hearing the rumors, Berg telephoned Cohen. "Stan," he said, "I understand you're thinking of writing a letter that might conflict with the letter my committee will be issuing soon at the request of the academy." "Yes, Paul," said Cohen, "I am. Your committee now has no one who knows about plasmids." "I think it would be silly to have two letters on the same subject," Berg said. "Why don't you join my committee and we'll work out phraseology to please us all."

Cohen agreed to join the committee. Of course everyone else at Stanford knew he had joined. And all the other hybrid molecule builders wanted to join too. Like Cohen, they felt they had interests to protect.

To keep peace among the Stanford family and its friends, Berg accepted three others along with Cohen. One was Professor Herbert Boyer, the University of California (San Francisco) Medical School enzyme specialist who had been Cohen's partner in the original plasmid DNA molecule experiment. One was Dr. Ronald Davis, the Stanford electron microscopist who had shared Janet Mertz's discovery of the unique nature of EcoRI enzyme. One was Stanford Professor David Hogness, who had worked with Professor Kaiser in the experiment that had introduced foreign DNA into *E. coli* for the first time, and who more recently had constructed hybrids of the DNA of eukaryote *Drosophila,* the fruit fly, with the DNA of prokaryote pSC101 and with the DNA of a bacteriophage.

15

THE CHANCES that any small group of "undersigned individuals," lacking institutional status and divided among themselves, could persuade the scientists of the world to voluntarily defer a line of experimentation that had better than usual prospects of yielding Nobel prizes were so vanishingly small as to seem nonexistent. Belonging to a family steeped in the intricate statistics of prescription drug research, manufacture, and marketing, Dr. Richard Roblin could not have been so naïve as to believe otherwise. Yet "undersigned individuals" was the only honest designation he could give to those who were to sign the open letter from the so-called Berg committee to the scientific community. And Roblin made that designation without displaying the smallest symptoms of stoicism. By the time he got word from Palo Alto that the list of signers would be four names longer than originally planned, he was well into the third draft of the letter, which had meanwhile passed through several metamorphoses.

"We wish to call attention to the rapidly increasing facility with which DNA molecules from diverse sources can be joined together," the first draft had said. "Personal communications from several groups of scientists indicate that experiments using this technology to create recombinant DNA from a variety of viral, animal, and bacterial sources are currently in progress or being planned. Although such experiments would facilitate the solution of many important biological questions, they would also create new types of infectious DNA elements which could prove dangerous to man."

The expressed concerns were traced back to the 1973 Gordon Research Conference on Nucleic Acids. The National Academy of Sciences request for advice was referred to, and the "undersigned individuals" pledged themselves not to participate in the questioned experiments until "attempts to evaluate the hazards have been made."

The director of the National Cancer Institute was urged to "give priority consideration to setting up an experimental program, under stringent biohazard containment conditions," to make the stipulated evaluations, and the convening of a "meeting of interested scientists ... no later than the spring of 1975" was proposed "to review scientific progress in this area and to further discuss appropriate ways to deal with the potential biohazards."

The second draft of the letter did not simply call attention to the problem. "We are concerned," it said. The appeal to the director of the National Cancer Institute was continued. The proposal for the conference was repeated, and a request was made that, pending the conference, "all scientists . . . join . . . the undersigned individuals . . . in voluntarily deferring" all experiments involving construction of hybrid DNA molecules that might spread cancer or create new strains of bacteria resistant to antibiotics. The proposed period of deferment was extended to encompass not only evaluation of hazards but achievement of "some resolution of the outstanding questions."

A heightened sense of urgency on the part of the "undersigned individuals" was reflected in the third draft of the letter. "We are seriously concerned," it said, "that some of these artificial recombinant DNA molecules could prove hazardous to man." The particular potential hazard that Robert Pollack had stressed in his private challenge to Berg in 1971—the same one that *Nature*'s cell biology correspondent had spelled out publicly in 1972—was formally identified: "the need to use a bacterium such as *E. coli* to clone the recombinant DNA molecules and to amplify their number. It was explained that *E. coli* strains commonly reside in the human intestinal tract, and they are capable of exchanging genetic information with other types of bacteria, some of which are pathogenic to man."

Official responsibility for the safety of hybrid DNA research was no longer thought to be containable within any one of the National Institutes of Health. It belonged at the very top of NIH in the hands of the director. Furthermore, the director should be guided in his decisions by an advisory committee of scientists charged with "overseeing an experimental program to evaluate the potential hazards of . . . DNA molecules," developing procedures that would minimize spread of the hazards "within the human population," and "devising guidelines to be followed by investigators" working with DNA molecules.

Roblin sent copies of the third draft of the letter to old and new members of the committee alike. An extra copy went to David Baltimore, who had earlier agreed to carry it by hand from Cambridge to Washington, D.C., with the intention of obtaining an informal "approval in principle" from the executive committee of the Assembly of Life Sciences of the National Academy of Sciences on May 24. By that time, Dr. Raymond Vosberg had replaced Leonard Laster as the Berg committee's liaison with the academy. On the day of the assembly meeting with Baltimore, Vosberg reported to Roblin by phone that the Berg committee's letter was acceptable to the assembly as it stood. Before that month of May was out, Nathan and Weissman with no difficulty cleared the text with selected members of the European Molecular Biology Organization at a meeting in Belgium, and the "undersigned individuals" entered a state of euphoria until . . .

On May 28, four days after the oral okay of the letter by the executive committee of the academy's Assembly of Life Sciences, Roblin received another phone call from Washington. It was Vosberg again. Academy President Philip Handler was having some second thoughts about the Berg letter. Written confirmation of those thoughts arrived in Cambridge a few days later, and the contents of that communication in turn were amplified in a face-to-face talk between Handler and Roblin at Minneapolis, where Roblin had gone to participate in the June 4 symposium that had prompted his inquiry to Maxine Singer back in February.

Baltimore meanwhile had walked the letter through an early June workshop at the Cold Spring Harbor Laboratory, where he began his presentation by asking if any newsmen were present. Hearing no response, he expressed the opinion that proponents of gene splicing should consistently emphasize the favorable elements in the situation.

Some of the younger scientists present told a science reporter they felt that such a course would amount to deliberately misrepresenting the facts to Congress and just as deliberately shortchanging the taxpayers. No one made this point aloud until afterward, however, and the silence left an illusion of unanimous concurrence with Baltimore's stated view.

At Minneapolis, Handler spent an hour making his unhappiness with the Berg committee letter plain to Roblin. The phrasing, the academy president said, left the insulting impression that the academy was serving Berg and company as a free post office. Handler could not allow that. The academy was being criticized for failing to take initiative in the public interest. And here was a matter of great public concern in which the academy had promptly appointed a study committee and had paid the committee's expenses; now the academy was about to be robbed of due credit. Handler wanted the Berg letter to clearly identify the Berg committee as a creature of the academy.

Berg was in Australia for a short visit just then. In his absence, Roblin felt uncomfortable. Laster had said, "Remember not to call your committee a committee of the academy." Now the president of the academy was saying: "You're an academy committee; don't forget it." As secretary of the little group of "undersigned individuals," Roblin had no authority to argue the case either way. He telephoned Berg in Australia and conveyed the gist of the talk with Handler. Berg was nettled by the academy's apparent inability to make up its mind. He asked Roblin to do his best to incorporate Handler's thoughts into a fourth draft of the letter. Roblin was willing. But his bosses at Harvard were complaining about the time he was spending away from his laboratory, and he adopted the timesaving expedient of putting a new paragraph at the top of the letter's third draft. That paragraph declared: "The following statement was prepared by a committee of the Assembly of Life Sciences of the National Research Council and is endorsed by the executive committee of the assembly." At the end of the letter, above the signatures of the committeemen, Roblin then inserted a formal designation for the group. "Ad Hoc Committee on Synthetic Nucleic Acids, Paul Berg, chairman."

When Berg returned from Australia, Roblin, still under time pressure to keep abreast of his lab work, asked to be relieved of further involvment in the letter drafting chore. Berg and Handler finished the fifth and final draft alone, quickly agreeing that their differences could be resolved by making three mechanically small but meaningful large revisions in the letter. The first revision removed the introductory paragraph that Roblin had grafted onto the third draft.

The second revision brought into a single paragraph the 1973 Gordon Research Conference on Nucleic Acids' request that the academy consider the possible dangers of hybrid DNA molecules, the academy's request to Berg to organize a study committee that would act on behalf of the academy's Assembly of Life Sciences, and the assembly's endorsement of the committee's recommendations. The third revision identified the eleven members of the Berg committee as the "Committee on Recombinant DNA Molecules, Assembly of Life Sciences, National Research Council, National Academy of Sciences."

With those finally determined facts duly entered, the fifth and last draft of the Berg committee's open letter on the potential promises and hazards of hybrid DNA molecules went to the academy's public information office for processing.

16

HOWARD J. LEWIS, the public information officer of the National Academy of Sciences, is one of the ablest living practitioners of his craft. He organized the office he now presides over in 1957, at the invitation of the late Dr. Detlev Bronk, who in those days was known as "Mr. American Science" in tribute to the wisdom with which he simultaneously headed the academy, the President's Science Advisory Committee, the governing board of the National Science Foundation, and Rockefeller University. Before joining the academy staff, Lewis wrote for newspapers and magazines and took with him to the academy a firm commitment to the rules that journalists everywhere respect—"Don't play favorites"; "Always respect a confidence"; "Reward individual initiative wherever you can; otherwise, break the news as fast as it happens no matter who gets it first."

To Lewis, academy committee reports and press conferences elucidating them were as inevitably related as cause and effect. The Berg committee's open letter clearly constituted a committee report, and a press conference was taken for granted.

The last minute word to Lewis, from President Handler's office, was that the Berg committee letter should go, under academy aegis, to three scientific journals simultaneously: *Science,* official organ of the American Association for the Advancement of Science; *Nature,* the British scientific bible; and the *Proceedings* of the academy. Of these three, the one that offered quickest publication was *Science.* Taking pains to alert *Nature* at the same time and obtain assurance that a page in its next edition would be held open for the news, Lewis personally delivered the letter by hand to Nicholas Wade, a member of the editorial staff of *Science,* who waited in his office until late in the day in order to hold the doors open for Lewis's arrival. Use of that express delivery system allowed the letter to be scheduled for publication in the July 26 issue of *Science.* As issues of that journal are dated a week later than the day copies are due to reach subscribers, the press conference was set for July 18.

When Lewis advised Berg of this calendar of events, however, Berg objected. He said he hadn't understood that a press conference was part of an academy committee's responsibility, and he had consequently agreed to a more exclusive arrangement.

On the Stanford campus at Palo Alto, Stanford public information officer Spyros Andreopoulos had no suspicion of the awkward episode at the academy. He was having a problem of his own. He had discovered the existence of the Berg committee letter by accident, and had asked Berg for a copy of it. When Berg complied, Andreopoulos wrote a press release based on the letter and took it to Berg.

"You can't send that out," Berg said. "Why can't I?" Andreopoulos asked. "Because we promised a twenty-four-hour exclusive to Victor McElheny, of the *New York Times.*"

"We?" asked Andreopoulos. "Who are we?"

"Not I," said Berg. "Baltimore did it. But we can't cross Dave up."

"I didn't promise anyone an exclusive," Andreopoulos said. "I have an obligation in the opposite direction—to see that every science writer on the West Coast gets the story at the same time. We count on these fellows to print our small news items. Now that we have a big one we can't let them be cheated of it."

Blood pressure soaring, Andreopoulos returned to his own office and telephoned Howard Lewis in Washington. Lewis, it turned out, had assigned the handling of the Berg letter to Brad Byers, his spot news chief. Andreopoulos talked to Byers. Byers had no background information on the Berg committee letter to offer to Washington reporters. Did Andreopoulos have anything? Not much, but something. And Andreopoulos teletyped what he had to the academy.

In the telephone exchanges that followed, Andreopoulos confirmed Berg's story about McElheny. Baltimore indeed had promised a twenty-four-hour beat to McElheny. Byers said Lewis didn't like the arrangement. It violated the academy's standard practice. But surrounding circumstances were such that the promise to McElheny could not be broken honorably. McElheny somehow had seen a copy of an early draft of the Berg letter. He had agreed to hold up publication in return for the limited "scoop." He had for some time abstained from releasing the story on the understanding that when the news did break he was entitled to an exclusive.

Lewis did his best to honor the promise to McElheny. However, there was no way to avoid the customary academy press conference. For the first time in his memory, Lewis had to try to interest Washington correspondents in a press conference without telling them what the conference would be about. They were much too busy and too canny a lot to fall for any such vagary. "What's it about?" they kept asking. Finally, fearful of losing his audience, Byers said: "It's an important development in biology." Most new developments in biology in one way or another involve the National Institutes of Health. That was clue enough for Stuart Auerbach, of the *Washington Post,* who began querying his NIH sources by phone. Within half an hour, he knew the origin of McElhenry's "scoop" and had the story himself. He told Lewis he was going to print it.

The *Times* published McElheny's version of the Berg committee's recommendations on the morning of July 18, the day set for the academy press conference. A two-column headline on the front page said: "Genetic Tests Renounced Over Possible Hazards." The lead paragraphs read:

> In an action rare in the history of science, prominent American biologists, including one winner of the Nobel Prize, are voluntarily renouncing for the present two types of genetic experiments that they consider could be hazardous.
>
> The classic approach in the sciences has been to pursue a trail of scientific inquiry wherever it may lead. However, in line with a growing scientific concern about some implications of modern molecular biology, the scientists warned against "indiscriminate application" of new techniques involved in the experiments.
>
> The scientists are announcing this action in a letter that, within the next week, will reach much of the world scientific community. They said they were taking the step because the gene-transplantation experiments might accidentally increase the resistance of some micro-organisms to drugs, or lead to the spread of some types of cancer-causing virus.

But the story was no longer a McElheny exclusive. Across the country, the *San Francisco Chronicle* carried a double-column front-page headline—"A Danger in 'Man-Made' Bacteria"—and beneath it a report from *Chronicle* science correspondent David Perlman, which began:

> A committee of eminent American scientists is urging colleagues around the world to halt some of their research aimed at modifying the genes of viruses and bacteria because their semi-artificial organisms

> might escape from the laboratory and endanger the general population.
> No one is yet sure what the hazards really are in creating "hybrid" genetic molecules that could infect higher organisms, including man.
> But because there is a remote possibility they might prove catastrophic, the committee members themselves have already imposed a moratorium on some of their own work in this area, even though it could eventually prove extremely valuable to mankind.
> The work arises from newly-discovered experimental techniques for creating semiartificial genetic molecules with enormous potential benefits: They could play a key role in cancer control, for example; or permit the mass production of vital biological substances such as insulin and growth hormone; or yield new strains of plants capable of flourishing without artificial fertilizer.
> But all this hope is now coupled with a warning:
> "There is serious concern that some of these artificial molecules could prove biologically hazardous."

Auerbach broke his story in the *Post* that same morning.

The barrage of premature broadsides did not prevent a good turnout for the academy press conference, which took place the midmorning of July 18. Copies of the full text of the open letter were distributed. Berg, Baltimore, and Roblin appeared to answer reporters' questions. Baltimore had prepared a statement in advance, with the approval of his two committee colleagues. It is not often referred to in accounts of the hybrid DNA molecule controversy, and I quote it in full here because of what it reveals about the Berg committee's position at that time.

> A recent series of advances in molecular biology has led to new methods for manipulating the molecules of heredity called DNA. Pieces of DNA from any source can now be caused to multiply inside a bacterium. By growing the bacterium, it is then possible to produce enormous amounts of the specific DNA of interest.
> This technique is rapidly being exploited in a number of laboratories to answer important outstanding questions in biology. The technique holds the promise of generating new ways of making therapeutic compounds such as insulin. It might also be used to modify bacteria so that new strains can be developed which could turn nitrogen from the atmosphere into plant food.
> While it is a crucial advance in the methodology of molecular biology, if the techniques were to be employed with certain types of DNA, the bacterium carrying this DNA could represent a danger.
> Two types of DNA which could prove hazardous are those which cause bacteria to be resistant to antibiotics or which cause the formation

of bacterial toxins. The other potentially hazardous form of DNA is any type of DNA which is part of, or derived from, a virus able to multiply in animals. A special worry are viruses which are able to cause cancer in animals.

In response to the concern of many members of the scientific community about the hazards implicit in this type of research, a committee of the National Academy of Sciences has issued a statement called "Potential Biohazards of Recombinant DNA Molecules." The statement outlines the potential hazards and binds the members of the committee not to utilize in their own experiments any of the types of DNA which are of concern. It also asks the rest of the scientific community, both in the United States and abroad, to join the committee in voluntarily deferring these types of experiments.

The committee report points out that a more remote possibility of hazard exists in other types of experiments and suggests that those experiments should be done only with the appropriate concern for the hazard.

The committee further asks the Director of the National Institutes of Health to establish an advisory committee to oversee the evaluation of hazards in the described experiments and to develop procedures which could allow these experiments to be carried out without the attendant hazard. The committee hopes that it will be possible to proceed with the experimentation in the near future when the potential for hazard has been overcome.

Finally, the committee proposes that an international meeting be held at which the scientific community from all over the world will try to assess the pace at which this type of research should proceed.

The technology required to carry out the experiments for which deferral has been proposed is quite simple and the time required to carry out the experiments would be short. Therefore, in order to reach the scientific community rapidly with its appeal, the committee has asked two weekly journals, *Science* and *Nature,* to publish the report. Both journals, along with the *Proceedings of the National Academy of Sciences, U.S.A.,* have agreed to carry the report in their next issue.

The committee wishes to emphasize that nothing which has been done thus far by scientists has produced any known hazard and that the risk of hazard is not enormous. Rather than accept the risk of hazard, however, the committee felt it more prudent to evaluate the hazard before carrying out the experiments.

Five points deserve particular attention. First, the scientists were dealing with a new technology, a method of domesticating bacteria to mass-produce genes and their products. Second, the Berg committee acknowledged that concern about the hazards was shared by "many." Third, the Berg group in its letter was speaking as a committee of the academy. Fourth, the committee did not expect the deferment of experiments to continue for very long. Fifth, the committee itself didn't think the risks were very great, but agreed that evaluation of the risks should precede the experiments.

The press conference lasted about an hour. As soon as it was over, Leon Jacobs, an NIH administrative official who had gone to the academy for the occasion, consulted with Berg, Baltimore, and Roblin concerning financial arrangements for the international conference at Asilomar early in 1975. Academy President Handler had already received a letter from NIH director Robert Stone telling of prior NIH interest in the establishment of guidelines for research. Although young Dr. Andrew Lewis, of the National Institute for Allergy and Infectious Diseases, was not mentioned, it was obviously his efforts to generate among scientists a deeper sense of responsibility to society that Stone was referring to. It is interesting to note that although the Berg committee had placed a government program of evaluating risk research first on its proposed agenda, all the emphasis from the very beginning of NIH involvement was going to formulation of guidelines to allow the questioned experiments to continue.

After Jacobs left the academy, Berg, Baltimore, and Roblin went to lunch and continued their discussion into the afternoon. They agreed that a somewhat augmented committee would meet again on September 10 to begin planning the Asilomar conference. The conference would revolve around the work of four study panels: one on the ecology of the bacterial workhorse *E. coli;* one on plasmids; one on viruses; and one on animal cells. A chairman would be chosen for each panel, and it would be up to him to assemble from across the world the most expert minds in his area of research into hybrid DNA molecule promises and problems.

17

THE NEWS and Comment section of the issue of *Science* that carried the Berg committee letter included a report that reflected the confusion generated by the flip-flop in attitude within the National Academy of Sciences. "Although endorsed by NAS," the *Science* item said, "the group's statement . . . is intended to be the personal appeal of the signatories." Later on the commentator reiterated: "The NAS endorsement gives the statement an even stronger right to a hearing, though does not in fact mean that the group is claiming to speak for anyone other than themselves." The academy's original reluctance to accept the committee wholeheartedly was wrongly assumed to persist at the time of the moratorium appeal's release and might, the reporter supposed, "mitigate criticisms that the group has too narrow a membership." As we shall see, this supposition was also wrong.

The *Science* news writer had undertaken a "preliminary tasting of opinion" among scientists easily available by telephone. It suggested that the "majority of scientists" would "firmly endorse . . . a temporary ban on Type I and Type II experiments," those involving cancer-provoking viruses and plasmids carrying bacterial resistance to antibiotics. (See page 97 for the complete text of the Berg committee letter as it was reproduced in *Science.)* However, the informal poll revealed that spirited argument had already developed over a third type of experiment specifically mentioned in the open letter. This type, the letter had said, "should not be undertaken lightly." It was the type that involved introduction of animal genes into bacteria.

"Virologist Wallace P. Rowe, of the National Institute of Allergy and Infectious Diseases, considers that such experiments should only be done when bacteria are found that are quite unable to infect man," the *Science* report said. It also quoted Robert G. Martin, chairman of the NIH biohazards committee, as advocating (personally rather than officially) "complete abstinence" from the third type of experiment.

On the opposite side of that issue *Science* found Donald R. Brown, of the Carnegie Institution of Washington, who had supplied the *Xenopus laevis* DNA that John Morrow, Stanley Cohen, and Herbert Boyer had replicated in *E. coli* and who had gone on to plan the hybridization of silkworm DNA in order to replicate the genes that govern the making of natural silk.

"I cannot see how this could be any conceivable danger to anybody," Brown was quoted as saying. Informed of one scientist who had expressed concern over "getting a gutful of silk," Brown replied that the differences between bacterial and animal cells were so great that the *E. coli* would reproduce only the silk-making genes and not the silk product that the genes control.

The other side of the coin that Brown had so casually flipped into the air was not examined. It would have reminded anyone who looked at it closely that if bacteria will not produce the products of animal genes, then the brightest promise that the proponents of gene splicing were holding forth to the public—fast, cheap manufacture of insulin and other vital hormones—could not be kept.

Science noted that most of the members of the Berg committee were "scientists who are already active in, or have considered entering, the research field in question." Thus, "at first appearance," the committee was "vulnerable to portrayal as the fox set to guard the henhouse." But "the group's unfoxlike recommendations are evidence to the contrary."

Was this entirely true? Had not *Science* overlooked the presence on the Berg committee of Dr. David Hogness, the Stanford professor who was strenuously pushing the splicing of DNA molecules of fruit flies, which are, of course, primitive animals? It was a glaring omission. For the news commentary itself characterized the promise of the fruit fly experiments as being "so great that those involved . . . are likely to resent any suggestion that . . . their plans . . . should be stopped or even postponed."

In those days of the first appearance of the Berg committee recommendations, *Science* reported that there was already abroad "fear that mere formalization of such proposals will lead to further impediments" in the path of research and to removal of the "ultimate decision from the hands of scientists." Why anyone should ever have thought that the "ultimate decision" in a democratic society belonged in the hands of scientists was not explained. Nobelist Joshua Lederberg was quoted as saying that the "ultimate purpose [of the Berg committee letter] is excellent but there is already such a momentum toward the regulation of research that the proponents should carefully consider the consequences of stating such recommendations." *Science* fairly could have but did not recall that it was Lederberg above all others who argued the necessity for the National Aeronautics and Space Administration to construct an isolation station for the debugging of astronauts and the rocks they brought home from the moon. However, Lederberg's permissive attitude toward scientific hunting privileges was balanced against the "quite contrary view" of Harvard's Jonathan Beckwith, who had spoken out on the issue in regard to his own work in 1966 and now in July 1974 was expressing himself as "happy to see this [Berg committee] precedent set because it will raise a debate about academic freedom to pursue whatever research one wishes."

Science's commentary on the Berg committee letter ended with this statement of opinion: "Quite possibly the embargo [against the two designated types of research] will be observed until the conference [called for Asilomar] in February. Its real test will come when and if the conference decides the hazard is substantial enough for the embargo to be indefinitely extended. It could then become apparent that control of the new technique is not much easier than the containment of nuclear weapons."

That statement took on something of a doomsday cast when read alongside a prediction made by an unidentified NIH scientist. "Anyone who wants to will go ahead and do it," this scientist asserted, adding that gene splicing would be "a high-school project within a few years."

Potential Biohazards of Recombinant DNA Molecules

Recent advances in techniques for the isolation and rejoining of segments of DNA now permit construction of biologically active recombinant DNA molecules in vitro. For example, DNA restriction endonucleases, which generate DNA fragments containing cohesive ends especially suitable for rejoining, have been used to create new types of biologically functional bacterial plasmids carrying antibiotic resistance markers *(1)* and to link *Xenopus laevis* ribosomal DNA to DNA from a bacterial plasmid. This latter recombinant plasmid has been shown to replicate stably in *Escherichia coli* where it synthesizes RNA that is complementary to *X. laevis* ribosomal DNA *(2)*. Similarly, segments of *Drosophila* chromosomal DNA have been incorporated into both plasmid and bacteriophage DNA's to yield hybrid molecules that can infect and replicate in *E. coli (3)*.

Several groups of scientists are now planning to use this technology to create recombinant DNA's from a variety of other viral, animal, and bacterial sources. Although such experiments are likely to facilitate the solution of important theoretical and practical biological problems, they would also result in the creation of novel types of infectious DNA elements whose biological properties cannot be completely predicted in advance.

There is serious concern that some of these artificial recombinant DNA molecules could prove biologically hazardous. One potential hazard in current experiments derives from the need to use a bacterium like *E. coli* to clone the recombinant DNA molecules and to amplify their number. Strains of *E. coli* commonly reside in the human intestinal tract, and they are capable of exchanging genetic information with other types of bacteria, some of which are pathogenic to man. Thus, new DNA elements introduced into *E. coli* might possibly become widely disseminated among human, bacterial, plant, or animal populations with unpredictable effects.

Concern for these emerging capabilities was raised by scientists attending the 1973 Gordon Research Conference on Nucleic Acids *(4)*, who requested that the National Academy of Sciences give consideration to these matters. The undersigned members of a committee, acting on behalf of and with the endorsement of the Assembly of Life Sciences of the National Research Council on this matter, propose the following recommendations.

First, and most important, that until the potential hazards of such recombinant DNA molecules have been better evaluated or until adequate methods are developed for preventing their spread, scientists throughout the world join with the members of this committee in voluntarily deferring the following types of experiments.

Type 1: Construction of new, autonomously replicating bacterial plasmids that might result in the introduction of genetic determinants for antibiotic resistance or bacterial toxin formation into bacterial strains that do not at present carry such determinants; or construction of new bacterial plasmids containing combinations of resistance to clinically useful antibiotics unless plasmids containing such combinations of antibiotic resistance determinants already exist in nature.

Type 2: Linkage of all or segments of the DNA's from oncogenic or other animal viruses to autonomously replicating DNA elements such as bacterial plasmids or other viral DNA's. Such recombinant DNA molecules might be more easily disseminated to bacterial populations in humans and other species, and thus possibly increase the incidence of cancer or other diseases.

Second, plans to link fragments of animal DNA's to bacterial plasmid DNA or bacteriophage DNA should be carefully weighed in light of the fact that many types of animal cell DNA's contain sequences common to RNA tumor viruses. Since joining of any foreign DNA to a DNA replication system creates new recombinant DNA molecules whose biological properties cannot be predicted with certainty, such experiments should not be undertaken lightly.

Third, the director of the National Institutes of Health is requested to give immediate consideration to establishing an advisory committee charged with (i) overseeing an experimental program to evaluate the potential biological and ecological hazards of the above types of recombinant DNA molecules; (ii) developing procedures which will minimize the spread of such molecules within human and other populations; and (iii) devising guidelines to be followed by investigators working with potentially hazardous recombinant DNA molecules.

Fourth, an international meeting of involved scientists from all over the world should be convened early in the coming year to review scientific progress in this area and to further discuss appropriate ways to deal with the potential biohazards of recombinant DNA molecules.

The above recommendations are made with the realization (i) that our concern is based on judgments of potential rather than demonstrated risk since there are few available experimental data on the hazards of such DNA molecules and (ii) that adherence to our major recommendations will entail postponement or possibly abandonment of certain types of scientifically worthwhile experiments. Moreover, we are aware of many theoretical and practical difficulties involved in evaluating the human hazards of such recombinant DNA molecules.

Nonetheless, our concern for the possible unfortunate consequences of indiscriminate application of these techniques motivates us to urge all scientists working in this area to join us in agreeing not to initiate experiments of types 1 and 2 above until attempts have been made to evaluate the hazards and some resolution of the outstanding questions has been achieved.

PAUL BERG, *Chairman*
DAVID BALTIMORE
HERBERT W. BOYER
STANLEY N. COHEN
RONALD W. DAVIS
DAVID S. HOGNESS
DANIEL NATHANS
RICHARD ROBLIN
JAMES D. WATSON
SHERMAN WEISSMAN
NORTON D. ZINDER

Committee on Recombinant DNA Molecules, Assembly of Life Sciences.
National Research Council, National Academy of Sciences.
Washington, D.C. 20418

18

A SUDDEN scare crossed Professor Berg's horizon during the week following publication of the Berg committee letter. But the fright did not originate in a high-school lab prank. It came in a reported defiance of the plea for deferment of hybrid DNA molecule experiments that might spread virulent disease. The source of the bad news was a University of California student who had just returned to his home campus from a visit to the University of Washington at Seattle. The student said that Professor Stanley Falkow was using *E. coli* to replicate hybrids of the genes responsible for cholera.

Falkow is one of the foremost living authorities on *E. coli* and its plasmids. His opinions are respected by scientists everywhere. To have him openly flout the moratorium would be a smashing setback.

Had the student seen Falkow himself? No, Falkow was away from his lab at the time of the student's visit. Had the student inquired as to Falkow's whereabouts? Yes, Falkow was attending a conference somewhere on the Atlantic coast.

Facing a situation that he felt called for emergency action, Berg in Palo Alto telephoned Baltimore in Cambridge. Baltimore in turn called Richard Novick at the Public Health Research Institute of the

City of New York. Novick was chairman of the plasmid nomenclature panel that had been appointed at the Japanese-American plasmid conference at Honolulu in November 1972. Falkow belonged to that panel. So Novick might know Falkow's whereabouts. And Novick very soon did know, for Falkow walked into Novick's lab on his way home from New England, where he had been attending a Gordon Research Conference on Toxins and had given a progress report on his latest work.

Falkow has an unusual background. Son of first generation Jewish emigrants from Russia, he was born in Albany, New York, where both parents worked in the clothing trades, the father in shoes, the mother in corsets. Falkow was propelled into a scientific career by the traditional Jewish interest in the professions and coincidentally by his reading of Paul de Kruif's book *The Microbe Hunters* at the age of twelve. From that time onward, he was determined "to be like Pasteur and Koch." He could not afford medical school, but it cost nothing to make friends with hometown hospital personnel, and by the age of twenty he was observing autopsies on the bodies of victims of infectious diseases. The emphasis of that last sentence belongs on the last four words. An urgent need to discover why pathogenic (disease-causing) organisms are pathogenic has dominated Falkow's life. He qualified early as a medical technician, and had the good fortune to study under the great Allen Campbell at the University of Michigan before being knocked out of action for a long while by a virulent attack of hepatitis. After recovering from that derailment, he obtained National Cancer Institute funds for Ph.D. studies at Brown University.

A few of his exploits in plasmid research were mentioned earlier in these pages. But his largest contribution to the comfort of the greatest number of people undoubtedly will come from his work with the *E. coli* gene that controls diarrhea in warm-blooded animals, including man.

To most of us, diarrhea is a momentary inconvenience, usually lasting no more than a few days. But it can threaten life if loss of body water becomes excessive. Particularly susceptible are infants and older people with cardiovascular problems, which become complicated when the vital electrolyte balance, governing the body's electrical impulses, is disturbed by shifts in fluid volume. Severe diarrhea can trigger heart attacks. The same internal mechanisms that bring on diarrhea in man also operate in animals, where the loss is far greater. A vaccine against diarrhea would be a boon to human travelers, but a special blessing to farmers and livestock growers.

Research into diarrhea began years ago with the study of sick animals. The *E. coli* bacterium was a natural suspect. Particular strains of *E. coli* were isolated and their active agents filtered. The filtrates were injected into rabbits, whose intestines were surgically segmented under anesthesia. Some segments remained normal following injection of the filtrates. Other segments swelled like sausage links. The contents of the sausagelike segments were analyzed and the infection was traced from cell to cell in the laboratory. During Dr. H. William Smith's pursuit of that tedious chore in England in the late 1960s, Falkow was helping an Englishwoman scientist, Dr. Naomi Datta, to classify plasmids for an international catalog. He obtained some plasmid-carrying strains of *E. coli* from Smith through one of Smith's research associates, a Canadian veterinarian, Dr. Carlton Gyles, who collected plasmids from stricken calves and pigs. After months of ransacking those plasmids Falkow found one in a strain of *E. coli* K12 that had the diarrhea toxin control gene in its DNA. He called that plasmid ENT, for enterotoxin (a poisonous agent in the intestine). Between 1968 and 1972, he recovered ENTs from many specimens of *E. coli* and moved them with him from Washington, D.C., to Seattle when he left the Georgetown University faculty to join the faculty of the University of Washington.

It was in November 1972, it will be recalled, that he took part in the now famous bargain that was struck between Professors Stanley Cohen and Herbert Boyer at the kosher delicatessen on Waikiki Beach in Honolulu. In return for supplying his plasmid RSF1010 to be cleaved by the restriction enzyme EcoRI and joined to Cohen's plasmid pSC101, Falkow would get both pSC101 and EcoRI for his manipulations of ENT.

At the time of that bargain, Falkow was being assisted in his laboratory work with ENT by a graduate student named Madelene So. In early 1973, as soon as he heard that the original splicing of the DNA molecules pSC101 and RSF1010 had been successfully achieved, he sent Mrs. So to Cohen's lab at Stanford to learn how to use pSC101 and EcoRI to deal with ENT. ENT contained many more genes than the toxin gene, and her task was to somehow whittle that one gene out of the DNA encasing it. Mrs. So found the competitive spirit of Stanford too fierce to enjoy and moved on to Boyer's lab in San Francisco to continue her whittling.

By the time the Berg committee began its deliberations, Mrs. So had used the restriction enzyme EcoRI to chop ENT into seventeen pieces. Originally, ENT had weighed 65 million daltons, the dalton being a standard measure of molecular size—one-sixteenth of the mass of an oxygen atom. Mrs. So put these pieces one by one into the pSC101 plasmid and then put the plasmid into *E. coli* to reproduce. After examining seventy-two different colonies of bacterial progeny, she detected the presence of the toxin gene in a piece that weighed only 5 million daltons. Next she made use of the fact that different restriction enzymes cut a given DNA molecule at different places, depending on the sequence of the genetic alphabet bases (the familiar As, Cs, Gs, and Ts), which each enzyme recognizes and attacks. Using roughly a half dozen of these enzymes in succession, Mrs. So cut the 5 million dalton piece of ENT down to an .8 million dalton fragment that still contained the toxin gene. Anticipating isolation of the pure toxin gene before long, Falkow decided it would be appropriate for him to report Mrs. So's progress at the 1974 Gordon Research Conference on Toxins late in July, during the period when release of the Berg committee letter was expected. With any luck at all, there was an excellent chance for a vaccine against diarrhea.

Having heard rumors of squabbling within the Berg committee, Falkow thought it wise to take precautions against any misunderstanding of his work with ENT. He wrote a letter to the disbursers of his federal government support funds, stating his intentions and requesting explicit approval for his plans. He got a clear go-ahead.

It was with considerable surprise and indignation, then, that Falkow heard from Novick that Berg and Baltimore had passed on the University of California student's story that Falkow was using *E. coli* to hybridize cholera genes. It was too much to take from people who had assumed the privilege of calling for a moratorium on experiments whose context they were very largely ignorant of. He telephoned both Baltimore and Berg and demanded an explanation. All his life he had followed the meticulous laboratory rules of medical microbiology, only to be accused now of adopting the sloppy practices of the newcoming molecular biologists. What possible justification could there be?

"If you were up on the subject you'd know that I'm working with the enterotoxin gene, not the cholera gene," he stormed. "I'm trying to isolate the toxin gene of *E. coli.* I'm cutting down a DNA fragment and am very close to the gene alone. I don't know how anyone who really understands what I am doing can believe I'm violating your moratorium. I'm not adding anything new to *E. coli.* I'm simply putting back into the bacterium a gene that has been there all the while."

"I'm glad you called, Stan," Berg said, in the characteristically disarming way that his intimates admire and strangers are baffled by. The fact that Falkow didn't intend to flout the moratorium made his criticism almost pleasurable to hear. "You're just the man I've been looking for. I'd like you to organize the opening session at the Asilomar conference next February on the ecology of *E. coli* and other enteric organisms and their plasmids. The session will include a panel discussion on *E. coli.*"

"There aren't enough scientists in this country who care about *E. coli* to make a good panel," Falkow replied. "I'd have to go abroad to get people."

"Okay," said Berg. "Go abroad for them."

19

WHEN PROFESSOR Roy Curtiss III, of the microbiology department of the University of Alabama Medical Center in Birmingham, read the Berg committee's moratorium plea in his office copy of *Science,* he was already engaged in a hybrid DNA molecule experiment that he hoped might eventually lead to eradication of tooth decay. He thought he had weighed all the possible risks before starting the experiment, but the content of the Berg committee letter brought him up short. He hadn't fully considered the range of biological hazards that might lie in expansion of the ecological niche occupied by the already ubiquitous *E. coli* bacterium. *E. coli* inhabits the big bowel and urinary tract and sometimes the upper respiratory passages of the human animal. What might be the consequence of its moving into the mouth as well?

Curtiss decided right there to hold up his experiment with dental caries until the potential problems could be thoroughly examined and resolved. Only in that way, he believed, could he as a scientist deserve the confidence of the people who were paying his research bills through their taxes. If other scientists did not share his definition of responsibility, he was willing to try to persuade them by citing instances in the recent past where science had carelessly forfeited clear opportunities to strengthen the public trust.

Unlike Professor Falkow, who had criticized the Berg committee letter and the limitations of its signers privately—on the telephone, to Berg alone—Curtiss wrote what he had to say in a single-spaced typewritten letter sixteen pages long, a well-constructed, smoothly flowing essay. On August 6, 1974, Curtiss mailed copies of that opus to more than a thousand scientists around the world as well as to all the members of the Berg committee including Berg himself. Some who didn't know Curtiss well but who got his letter thought he must have been very angry to put himself to all that trouble.

Curtiss was not angry. He was simply expressing some genes he had inherited from his paternal grandfather, the first Roy Curtiss, a New Yorker who had piled up a fortune as an advertising man and then lost it all in the great stock market crash of 1929 with massive debts to spare. The second Roy Curtiss was much less spectacular than his father but a more prudent manager. On the relatively low income of an accountant for the New York State Department of Public Welfare, Roy Curtiss II provided Roy III with a reasonably comfortable setting in which to practice grandfather's flair for promoting himself and his enterprises.

Most small boys who are lucky enough to win a baby duckling at a local firehouse bazaar keep the duckling until it gets sick and dies. Then they bury the bird with as much ceremony as family circumstances allow. When Roy Curtiss III won the firemen's prize duckling, he bought another duckling with the intention of starting a duck farm. He waited quite a while for one of the ducks to lay some eggs. When neither did, he consulted the person from whom he had bought the second duckling. How can a duck buyer distinguish a substandard duck? The answer taught him the importance of gender in genetics.

Having previously grown fifteen-foot sunflowers from seeds planted in an abandoned pony shed, Curtiss was accustomed to thinking big, even when the source of his modest prosperity was quite plainly fertilizer. So, after his brief frustration with the ducks, he turned his attention to chickens, raising them in batches of five hundred to six hundred, and winning a Chicken of Tomorrow contest by cross breeding broad-breasted Cornish hens and white Plymouth Rocks.

The time came for college. The agricultural school of Cornell University was not only well reputed but was located within easy travel distance of a county farm on which the family was then living on Long Island. But Cornell didn't turn out to be the seedbed of excitement Curtiss sought for his talents, and he took a two-year assignment as a technician at the Brookhaven National Laboratory of the Atomic Energy Commission. One of his research associates there was Myron Levine, who had come from Professor Salvador Luria's lab. Levine was experimenting with the P22 phage of the bacterium *Salmonella typhimurium.* P22 is the virus Peter Lobban would later use in DNA splicing. Through this accident of timing, Curtiss reached a new plateau in his education as a geneticist. He didn't previously realize that bacteria had their own viruses. He deepened his understanding of that symbiotic relationship by attending the yearly summer phage meeting initiated years before by Delbruck, Hershey, and Luria at Cold Spring Harbor Laboratory, thirty-five miles from Brookhaven.

One phase of his research at Brookhaven involved Curtiss in an important advance in molecular study. At that time the only radioactive tracers that DNA could be tagged with were carbon and phosphorus, which were difficult to work with, each for its own reasons. Curtiss was one of perhaps a dozen scientists who participated in experiments made possible by Peter Hughes's inventive addition of tritium—the three-atomed form of hydrogen—to thymidine, a compound form of the genetic alphabet's letter T. Because tritium releases its breakdown energy within a very short distance of its own location on photographic plates, the precision of DNA tracking was proportionately refined.

Just as the father of atomic power, Enrico Fermi, had made a nuclear furnace work for the first time at the University of Chicago, so had Leo Szilard there confounded his associates by applying physics to such seemingly distant biological problems as artifical limbs. In the footsteps of Szilard, Curtiss went in quest of a Ph.D degree. Unconsciously preparing for the revolution in molecular biology that would sweep him up ten years later, he concentrated on the genetics of the K12 strain of *E. coli.* With a doctor's certificate in hand, he moved again, in 1963, to the Oak Ridge National Laboratory of the Atomic Energy Commission in Tennessee. There he landed next door to the man whose work would lead Curtiss into the most exciting experiments of his life.

The scientist in the lab next door to Curtiss's lab at Oak Ridge was Howard Adler. In 1966 Adler described in detail a phenomenon that had been noticed occasionally since 1930. He called it a minicell. Minicells are progeny of bacteria that reproduce more often than they should. Minicells therefore are pinched off before they acquire the normal complement of chromosomal DNA. But if the parent bacterium carries a plasmid, daughter minicells contain that plasmid. Minicells can be separated from their normal, larger sisters in the centrifuge. Because of the simplicity of their structure, the minicells allow scientists to observe the action of RNA in directing the ribosomes to make proteins. In the minicells the synthesis of as many as five proteins at a time can be detected easily. In the minicells' normal sisters, the chromosome generates too much genetic noise to allow the signals of the proteins to be distinguished.

Having no idea what the minicells would mean to his future, Curtiss moved once more, in 1972, to the Medical Center of the University of Alabama in Birmingham. In the years immediately before that move, however, he added a new dimension to his life by buying a vacant lot in Oak Ridge and designing a house to put on it. Because he was his own architect, he made frequent inspection trips to the building site during construction of the house. On those trips his olfactory nerves were repeatedly offended. A quick reconnaissance located the source of the odor in a malfunctioning sewage disposal plant. By circulating a petition among neighbors affected by the nuisance, he pressured the city into fixing the plant. Not long after that, a seat on the Oak Ridge City Council opened and Curtiss's new reputation as a civic activist pushed him into the balloting arena. He came out of the election a 2-to-1 winner.

The busier he became as a councilman, the more he learned about the shaping of public opinion. A politician, he discovered, must listen to his constituents regardless of how few academic degrees they may have after their names. The two constituents he had to listen to most were his eldest sons. In 1970 they were thirteen and ten years old, respectively. Though proud of their father's influence, they felt it could and ought to be enlarged. He should represent youthful as well as adult thought, they believed, and to symbolize that purpose he (having aged within four years of the forty mark) needed to adopt some youthful customs. Long hair, for instance. They bet him $10 he wouldn't be able to live through the remaining ten months of the year without getting a haircut.

Curtiss not only won the bet but enjoyed the symbol of freedom immensely. But it was only a beginning. His work took him to Chile for a brief spell, and for some reason he cannot now explain he returned home determined to stop shaving. Curtiss enjoyed that freedom too. As his beard luxuriated, he felt no need to wear neckties. By the time of his appointment in 1972 as a professor of microbiology in Birmingham, he was an unforgettable figure, tall and broad shouldered, with a rugged face curtained by a thinning mane of dark brown hair that fell to and beyond his shoulders (except in summer, when he wears it in a pigtail), and a frizzled beard that flowed down his front—a natural for any movie-casting scout looking to recruit an Old Testament patriarch.

Curtiss continued to be true to his grandfather's genes when the moment came, in 1973, to decide by what route he should enter the promised land of hybrid DNA molecules. It is an advertising tenet to always have a dependable product to sell. An excellent product was close at hand for Curtiss—the research program of the University of Alabama Institute of Dental Research. It arose naturally in the wake of a former university president's role in discovering fluoride's importance in resisting dental caries. The institute today is one of the world's finest of its kind. Its work could hardly be closer to the daily needs of the people, for tooth decay is a universal affliction of the human animal and one of the most costly of all human infections.

As Curtiss told Berg in the sixteen-page letter of response to the moratorium plea of the Berg committee, the principal causative agent of dental caries is the bacterium *Streptococcus mutans (S. mutans* for short). Recent research has shown that *S. mutans* makes teeth susceptible to decay by manufacturing an insoluble enzyme that breaks sugar down into an acid that cuts through the tooth enamel and a sticky substance that holds undigested food particles to the spot where the enamel is broken. When *S. mutans* loses the ability to make this enzyme, the bacterium is no longer virulent. The manufacturing process is thought to be either directed or controlled by a plasmid inside the bacterium.

To discover the detailed mechanics of the plasmid's operation, Curtiss and a small group of associates had bred some of the minicells he had learned to work with at Oak Ridge. Simple transfer of *S. mutans* plasmids to *E. coli* minicells was attempted, but it didn't work. Curtiss was considering use of the restriction enzyme EcoRI to construct a hybrid of the *S. mutans* plasmid with an *E. coli* K12 plasmid that would be acceptable to *E. coli* when the Berg committee letter appeared.

"I personally pledge to cease Type I experiments [to construct bacterial plasmids that are not now known to exist] that I was currently engaged in," he wrote. "While the goals of such research are seemingly beneficial, the potential of endowing *E. coli* with the ability to occupy a new ecological niche and indeed to become cariogenic is somewhat frightening. In reality, however, the greatest fear that I perceived prior to abandoning this line of thought was the fact that such a microbe might also produce other streptococcal proteins which might, upon repeated exposure to humans, induce . . . complications associated with streptococcal infections."

"In order to be complete," Curtiss added, "I should also mention some additional types of genetic engineering experiments that we considered as potential means to displace virulent *S. mutans* from its normal ecological niche and thus to reduce dental caries. One line of attack would be to couple a bacterial gene coding for dextranase [an enzyme that dissolves the product made by *S. mutans*] to a plasmid present in a 'normal harmless' bacterial inhabitant of the oral cavity. A strain of bacteria that produced copious quantities of extracellular dextranase could then be selected, introduced into the oral cavity, and there act to enzymatically destroy the dextran-like polysaccharide and thus inhibit and/or prevent colonization of the tooth surfaces by virulent *S. mutans.*"

"A second, more far-out, approach" he had thought of was to couple to plasmid DNA some eukaryotic DNA sequences coding for blood fractions that would immunize a person against infection by *S. mutans,* a harmless bacterial inhabitant of the mouth, and finally to use the transformed bacterium to kill off *S. mutans* where it lives.

"I am well aware that the use of such 'microbiological warfare agents' would initially upset the ecological balance to the detriment of *S. mutans,*" Curtiss wrote. But, he said, the inherent genetic instability of DNA and the imposed selection pressures would eventually result in a new ecological balance that could even produce abnormally high numbers of variant, yet virulent, *S. mutans.*

The eventual experimental testing of such possibilities under very carefully controlled circumstances seemed warranted in his opinion "in view of the lack of existing effective and theoretically possible means to curb infection by this pathogen."

Essentially, Curtiss's letter proposed a much broader moratorium than the one the Berg committee had proposed—an almost total ban, in fact—not because he doubted the value of the experiments but rather because he felt their collective promise was so very great that scientists should clear up their own and other people's doubts with strictly contained tests so that the biohazards "can . . . be controlled" and "this important area of investigative research can continue for the betterment of all."

His analysis of the shortcomings of the moratorium as it stood was exhaustive. Bacterial plasmids had a greater range of powers than the Berg committee letter had suggested. Those powers could be transferred back and forth among at least thirty-three genera of bacteria. *E. coli* especially was grossly underestimated. Curtiss wrote:

> *E. coli* is often thought of as a harmless inhabitant of the intestinal tracts of animals. . . . *E. coli* is a pathogen with strains exhibiting various degrees of virulence. Indeed, infections with enteropathogenic strains of *E. coli* are probably responsible for the vast majority of diarrheal diseases and other enteric disorders among children and adults in the U.S.A. Furthermore, *E. coli* is one of the three main killers associated with patients dying of septicemias [blood poisoning] that secondarily arise because of diseases or states such as cancer, immune deficiency, transplantation, surgery, ulcers, appendicitis, peritonitis, etc. Consequently, infections due to virulent strains of *E. coli* result in significant economic losses and thus constitute one of our major medical problems.
>
> *E. coli's* status as a medically important pathogen is only now beginning to be realized . . . [because] *E. coli* infections were rarely encountered ten to twenty years ago except in cases of infantile diarrhea. Among *E. coli* strains isolated twenty years ago, plasmids were rather rare and when found usually were of but one kind. Today thirty to fifty percent of all *E. coli* strains harbor plasmids . . . with the mean number of molecularly distinct plasmids being about three per strain. Admittedly, the methodology of detecting plasmids has markedly been improved during this same interval of time. But I consider the increase to be very real and to represent a major evolutionarily significant change in the genetic potential of *E. coli* and of other microorganisms as well.

How did this evolutionary change originate? "Undoubtedly," Curtiss wrote, it was a "consequence of environmental changes that provide selection pressures [i.e., the use of antibiotics as feed additives for cattle, fish and poultry, etc.] and increased opportunities for conjugal plasmid and/or chromosome transfer" [i.e., due in part to increased levels of water pollution, etc.].

The extraordinary sexual prowess of the *E. coli* bacterium had to be taken into account too, the Curtiss letter said.

> I personally consider that *E. coli's* capabilities in these activities are probably unequaled by any other living organism (even bedbugs). *E. coli,* which normally divides every 20 to 30 minutes, can sustain the conjugal act for up to two to three hours, and most likely does so in the absence of vegetative chromosome replication and cell division. Mating partners are randomly selected without regard to size or shape, and multiple matings between a cell of one mating type and several of the other are a frequent occurrence in laboratory experiments at least. During conjugation, the donor genetic information is simultaneously replaced during its transfer to the recipient such that a donor cell is immediately able to engage in a second conjugal act following cessation of the first. . . . Unlike the infertility associated with homosexual couplings in other organisms . . ., homosexual matings in bacteria often lead to progeny containing recombinant plasmids and/or chromosomes. . . . *E. coli* also occasionally chooses its mating partners without regard to their taxonomic designations and is known to donate and/or receive plasmid and/or chromosomal information from members of over twenty genera. . . . Conjugal gene transfer has been demonstrated to occur in soil, the nodules on the roots of leguminous plants, and in the intestinal tracts of fish, poultry, rodents, cattle and humans. Plasmid and chromosome transfer also occur with equal frequency under both anaerobic [in the absence of oxygen] and aerobic [in the presence of oxygen]environments. Thus a great diversity of environments, some of which are "improving" because of greater pollution and/or microbial acquisition of plasmids that expand the ecological niches they occupy, are suitable for conjugal gene flow.

Nevertheless, Curtiss favored continuance of the use of *E. coli* as a molecular research vehicle, "even though I have some trepidation over this. At least a billion dollars has been spent by the countries of the world for research on this organism and without doubt we know more about *E. coli* than any other living organism. To not utilize this knowledge would be wasteful. Furthermore, the selection of some other microbe for these studies would be most difficult and potentially hazardous. . . . I see no advantages in using soil microorganisms or potential plant pathogens. In all instances severe, even though different, biohazards exist and there would be no lessening of the need for effective means to cope with these biohazards. Thus I believe *E. coli* to be the most suitable microorganism for these studies. Obviously, however, this matter requires further thought and debate."

Curtiss expressed his agreement with the Berg committee's recommendation for the establishment of an advisory committee by the director of the National Institutes of Health, but hoped "that an agency such as the World Health Organization would also become involved in facilitating discussions leading to an enumeration of the problems and the means to deal with them. In this way, the evaluation of the biohazards and the adoption of methods to deal with them would be the concern and responsibility of members of the entire worldwide scientific community."

Curtiss also concurred in the recommendation for an international meeting in 1975 to present new data and further discuss the associated problems. But he doubted that "such a meeting would be highly productive unless this topic is discussed and debated at other scientific meetings prior to that time and unless those scientists engaged or interested in this research and its ramifications begin to informally exchange views not only on the potential biohazards of this research but also on the means to effectively deal with these biohazards."

"Any given experimental situation will necessitate rather special guidelines," Curtiss thought, and "procedures to contend with any given biohazard cannot be universally applied. In addition to potential federal and/or international guidelines that may be established that would need to be complied with by scientists desiring to conduct such research, it might be well that scientists at public and private research institutions as well as at commercial firms that are interested or now engaged in this line of research have the advice, counsel, and approval of individual or institutional biohazards committees much as the human use committees are used to approve and supervise human research."

Regarding the training of personnel to participate in hybrid DNA molecule research, Curtiss's letter said:

> All persons engaged in this line of research, including scientific, technical and support personnel, should have appropriate training in conducting various biological, biochemical and containment procedures and should be well aware of all of the biohazards associated with the work and the means to contend with these biohazards. These

individuals should not only familiarize themselves with all of the written procedures, but should probably be certified as competent to be engaged in this research by an institutional biohazards committee.

Regarding the construction of hybrid DNA molecules:

> The cleavage of viral, plasmid and/or chromosomal DNAs with restriction endonucleases and their reassociation under favorable conditions to form recombinant DNA molecules should be done with great care. The left-over products of such reactions should be chemically, physically or enzymatically destroyed prior to disposal. Indeed, those investigators utilizing restriction endonucleases for elucidating the structure of DNA molecules should likewise take precautions to preclude introduction of such DNA fragments into the environment.

Regarding licensing of genetically altered organisms or their products:

> Licensing of the use of a product produced by a genetically altered organism or a preparation of the organism as a vaccine should adhere to the standards and requirements of licensing established by various federal agencies such as the Food and Drug Administration in the U.S.A. and equivalent licensing authorities in other countries. The introduction of a genetically altered organism into the ecosystem, however, should require conformance with requirements set forth by some international agency. In either case, data on efficacy and potential biohazards of the use of these products or organisms would need to be corroborated by more than one group of research scientists.

"This memorandum would not be complete," Curtis concluded, "unless I responded in some way to the general concerns about academic freedom and regulation of science that your initial letter elicited. By way of illustrating my own feelings on these matters, I need only remind others that scientists working in the area of radiation research and atomic energy did not heed the early warnings of the potential hazards and many suffered material damage before central agencies were established to place regulations on the conduct of this research and to protect the general public from radiation hazards. In this case, it seems that the scientists most knowledgeable about the dangers were altogether too slow in addressing themselves to the problems and in alleviating much of the pain and suffering which ensued. In the instances of regulation of human and/or animal research, again scientists were remiss in establishing guidelines for the conduct of their own experiments and, because of the few who did not adhere to the high standards of the majority, restrictive regulations were established that had to be adhered to by all."

In hybrid DNA molecule research, Curtiss saw an "opportunity to take a responsible stand that will establish the guidelines to permit this research to be conducted in a safe and beneficial way in the absence of unwarranted restrictions that would all but make this research impossible. Even though I disagree with you on some particulars I would thus like to take this opportunity to applaud you for your highly responsible and unselfish actions in initiating this dialog that will ultimately result in the safe resumption of this important area of research for the betterment of all."

20

IN THE April deliberations that preceded the drafting of its call for a moratorium on hybrid DNA molecule experiments, the Berg committee decided that one of its prime objectives—second only to the slowing or stopping of experiments suspected of high risk potential—would be to "make bad scenarios as difficult as possible by constructing plasmids and bacterial strains with built-in safeguards." Had a clear statement of that tenor found its way into the committee's open letter to the scientific community, it might have gone a long way toward minimizing or even perhaps eliminating the doubts and fears of many who weren't entirely sure of the motives of Berg and his associates. However, for no positive reason that anyone concerned can recall, the idea that the hazards of hybrid DNA experiments could be contained more effectively by biological than physical means was passed over entirely, not only in the text of the committee's letter of July but at the press conference announcing the issuance of the moratorium appeal.

The Berg committee met again on September 10 for the purpose of laying out the main structure of the program for the international conference set for Asilomar in February 1975. Six of the signers of the moratorium proposal (Professors Boyer, Cohen, Davis, Hogness, Nathans, and Watson) were dropped at that meeting to make room for Dr. Maxine Singer, co-chairperson of the 1973 Gordon Research Conference on Nucleic Acids; Dr. William Gartland, the federal government official NIH had chosen to be secretary of the yet unnamed Recombinant DNA Advisory Committee to NIH Director

Dr. Donald S. Fredrickson; and three discussion panel chairmen for Asilomar: Dr. Richard Novick, for the panel on plasmids; Dr. Aaron Shatkin, for the panel on viruses; and Dr. Donald Brown, for the panel on prokaryote-eukaryote hybrids. Each of the three chairmen was made responsible for naming his panel members, and Novick promptly telephoned Professor Roy Curtiss III to ask him to join the plasmid panel and expand the data contained in the sixteen-page letter Curtiss had broadcast on August 6.

Also at that September meeting of the Berg committee, two foreign scientists were nominated to serve with Berg, Baltimore, Roblin, and Singer on the Organizing Committee for Asilomar. One of the two was Dr. Niels Jerne, of the Basel Institute for Immunology in Switzerland (who was unable to accept), and the second was Dr. Sydney Brenner, of the MRC (Medical Research Council) Laboratory of Molecular Biology in Cambridge, England.

About two weeks after his appointment to the Berg committee was affirmed, Brenner presented evidence to the Working Party on the Practice of Genetic Manipulation, which the British Parliament had appointed in response to the moratorium appeal. In his testimony Brenner declared his belief that the "estimation of hazards should become part of the scientific evaluation of research projects" and that the evaluation should include the "much" that "can and should be done . . . to engineer the microorganisms used so that their chances of survival outside a laboratory environment are small."

As an example, he suggested the breeding of mutant bacteria that would no longer possess the normal ability to synthesize diaminopimelic acid (DAP). DAP is the raw material from which bacteria manufacture the stiff outer skin of their coats. The coats must grow in size, of course, as the bacteria grow. As long as DAP is available, the organisms can live comfortably. But when the DAP is gone, the insides of the bacteria reach beyond the outer limits of the coats and the bacteria literally explode. DAP can easily be supplied in the laboratory. As long as the bacteria remain in the laboratory, then they can survive. So far as has been determined, however, DAP is not present in large amounts in either of the refuges that would be natural to *E. coli:* the gut of warm-blooded animals, or sewage.

In giving his evidence at the British Parlimentary hearing, Brenner also suggested the construction of plasmids that cannot migrate from one bacterium to another, and the breeding of bacteria that are defective in transferring DNA sexually.

"This is clearly our responsiblity," Brenner said, "and we should shoulder it."

As a token of the seriousness he assigned to his membership on the Asilomar Organizing Committee, Brenner sent the above-quoted testimony to Berg, who passed it along to the Berg committee secretary, Roblin, for the committee files.

In October Professor Curtiss followed up his oral acceptance of Asilomar plasmid panel chairman Novick's telephoned invitation with a letter to another member of that same panel, Professor Falkow. The letter cataloged eleven specific ways in which microorganisms involved in the gene-splicing experiments could be disabled for life outside the laboratory. These included DAP deficiency, purine deficiency, carbohydrate deficiency, heat sensitivity, cold sensitivity, sensitivity to ultraviolet light, and sexual sterility.

But neither that direct prod from Curtiss nor the indirect one from Brenner moved the Berg committee to make a place on the Asilomar program for the novel concept of biological containment.

October passed, and November was almost gone when the British scientific journal *Nature* published a letter from one of its readers, J. H. Edwards, of the Queen Elizabeth Medical Center in Birmingham, England. Edwards wrote:

> Sir—
>
> The anxiety expressed by various scientists on the hazards of partially hybridizing certain types of microorganisms has already led to considerable public disagreement. It is clear that some responsible and established workers consider it unwise to take humanity with them on any tour of the unknown, while others consider it impractical to curtail the spread of organisms by any method proof against technical errors, psychoses, and even earthquakes. This will remain true whatever is stated by any of the many committees which are likely to advise each advanced nation.
>
> I wish to advance a simple positive proposal which cannot fail to increase the safety of any bacterial incorporation studies, and which would itself provide interesting examples of extreme adaptation which might have applications to other fields, such as waste disposal.
>
> Contracts should be offered for the development of an organism, with suitable qualities for work on partial hybridity, which would not grow within a pH range of 6–8 [the normal range of acid-alkali balance in a watery medium], would need an oxygen partial pressure of at least threefold the normal, and which was dependent on at least two unusual synthetic substrates. These should be very simple safeguards against any single mutation allowing reproduction in any natural host.
>
> Once such an organism was established, it should only be necessary to provide substantial subsidies to firms undertaking to provide the necessary culture media—and such media could hardly be misappropriated—to provide an economic and technical climate within which temptation to continue to work on the worst possible organism—a

commensal [tenant sharing the food] of the human gut—would be seen as both unnecessary and irresponsible.

Since even bacteria take time, and moratoria cannot be expected to contain curiosity for long unless alternative outlets are provided and seen to be imminent, the matter would seem urgent, and probably too urgent for the attention of the present public funding organizations. Even if no immediate funds are available from such sources, informed discussion might influence some companies which supply media and equipment for tissue culture to initiate work in the hope of reaping the large commercial rewards such developments could yield.

The Edwards' letter also went into the Berg committee file. But December came and the Asilomar invitations went out with nothing on the conference agenda to indicate that this highly imaginative approach to the laboratory biohazard problem would get any serious consideration whatever.

21

DANIEL SINGER, the public interest lawyer whose wife, Maxine, had co-chaired the 1973 Gordon Research Conference on Nucleic Acids, makes no secret of his attitude toward scientists "who view the public as hostile and somehow not worthy of their concern and attention." He deplores their "fancied omniscience," which, he says, can appropriately be described by the Yiddish word "chutzpa."

Singer therefore was pleased when Professor Paul Berg asked him in early November 1974 to help arrange press coverage of the forthcoming international conference on the potential hazards of hybrid DNA molecules. At the same time, Singer could not restrain his curiosity. He had been in on the planning for the conference from the moment his wife was named to the Organizing Committee. In fact, he headed a group known as "the Singer group," which included Leon Kass and Professor Alexander Capron, of the University of Pennsylvania. Singer, Kass, and Capron all were associated with the Institute of Society, Ethics and the Life Sciences. They were scouting for nonscientific participants in the meeting scheduled for the Asilomar Conference Center.

Why, Singer asked himself, had he, and not Howard Lewis, the public information officer of the National Academy of Sciences, been picked for this press assignment? After all, the committee of eleven who had called the conference was a special committee of the academy. Singer asked Berg whether there had been any discussion of the matter with Lewis. Yes, Berg said, he had spoken to Lewis and had been told that the academy preferred not to be involved in the type of press coverage that Berg had in mind.

Thinking that the path ahead was clear, Singer next consulted with his wife. As a member of the editorial board of *Science,* she was on a first name basis with the reporters on that journal's news staff. One reporter she knew especially well was Barbara Culliton. But if Maxine were to approach Culliton for advice, the move might be interpreted as obligatory. To eliminate that possibility, Daniel Singer went to see Culliton himself. He explained to her Berg's strong feeling that the number of reporters admitted to the conference should be limited. The number he had in mind was eight: three foreign (one representing radio, one TV, one newspaper) and five domestic (one radio, one TV, two newspapers, and one magazine). Berg was willing to invite that many with the understanding that all of them would sit through the entire meeting, which would probably last three to four days, and would file no reports to their publishers until after the conference broke up.

Did Culliton think such a system, which was sometimes used in Europe, would work in the United States, where the press is freer than anywhere else on earth? Culliton wasn't sure. But she was willing to ask a few of the top science writers in Washington and tell Singer what she found out.

Within a few days, Culliton reported back that some writers would accept the limitations if they could persuade their editors to agree. Singer in turn told Berg that the chances looked good and on November 8 met with Roblin to discuss what kinds of information could be made available to the attending reporters as the conference went along.

Because the scheduled meeting was still many weeks away, Howard Lewis, busy with other academy responsibilities, meanwhile gave only passing thought to Asilomar until a phone call came in from Stuart Auerbach of the *Washington Post.* Mindful of the circumstances that had scrambled publication of the Berg committee's moratorium appeal, Auerbach had given Berg early warning of the *Post's* continuing interest in the hybrid DNA molecule controversy. Specifically how, Auerbach wanted to know, should the *Post* go about arranging its coverage of Asilomar? Berg told him to ask Dan Singer. Instead of asking Singer, Auerbach asked Lewis: "Why must I clear my request to attend the Asilomar conference with Dan Singer?" Lewis didn't know, and said so. That didn't sound like Lewis. Auerbach got suspicious. When he pressed Lewis for an answer and couldn't get it, he threatened to invoke the Freedom of

Information Act in a court action if necessary to assure the *Post* of a seat at the Asilomar press table.

At least as annoyed as Auerbach had been, Lewis made an appointment to visit Singer at Singer's office. There Singer related all he knew about the situation. A relatively even-tempered man, Lewis doesn't often show anger. But he boiled over on Singer. Singer was hurt that his good intentions should be so thoroughly misunderstood. Hadn't Lewis told Berg that the academy wouldn't be interested in arranging the kind of press coverage Berg wanted for Asilomar? Yes, Lewis said, he had. But he hadn't meant to suggest that Berg or any other academy committee chairman was free to set up press relationships contrary to academy custom. Lewis had meant to say simply, and thought he had said very plainly, that the academy would insist on following the usual academy rule.

The flaming tempers of Lewis and Singer were effectively damped by the time the two men parted. But the fat was in the fire, as it had been when Lewis first learned of Victor McElheny's promised DNA "scoop" months earlier. Again Lewis was in the position of trying to work out a reasonable compromise that would save face all around.

As a first step, he sought to learn what practical reason there might be for limiting the number of reporters at Asilomar. He was told the Asilomar chapel was too small to accommodate all the scientists who were to be invited plus any sizable number of press representatives. Another source of information, however, made Lewis aware for the first time that the Asilomar chapel was in a large California state park. As any alert reporter would, Lewis sent for some travel brochures that would set forth the park's advantages to visitors who wished to use its facilities. The brochures convinced him there was plenty of room for a large number of reporters.

To uninformed outsiders, the Auerbach threat of a legal suit seemed to be a hostile attack on Lewis, for the suit would have to be brought against the academy in an area for which Lewis was plainly responsible. In practice, however, the threat proved to be a friendly and helpful act. Lewis was able to use it as a lever to push the limit on the number of reporters from eight to twelve and finally to sixteen. The official report that the Organizing Committee later submitted to the academy appended the names of only those sixteen: George Alexander, of the *Los Angeles Times;* Stuart Auerbach, of the *Washington Post;* Jerry Bishop, of the *Wall Street Journal;* Graham Chedd, of the *New Scientist;* Robert Cooke, of the *Boston Globe;* Rainer Flohl, of *Frankfurter Allgemeine;* Gail McBride, of the *Journal of the American Medical Association;* Victor McElheny, of the *New York Times;* Colin Norman, of *Nature;* Dave Perlman, of the *San Francisco Chronicle;*

Judy Randal, of the *Washington Star-News;* Michael Rogers, of *Rolling Stone;* Christine Russell, of *Bioscience;* Nicholas Wade, of *Science;* Janet Weinberg, of *Science News;* and Dermot A. O'Sullivan, of *Chemical and Engineering News.* But the text of the report to the academy said twenty-one reporters were in attendance. The discrepancy is due to events that no one expected.

If the Asilomar chapel on the morning of the conference opening on February 24, 1975, was not actually dominated by a paranoid fear of exposure to the public gaze, the introductory remarks and attitudes of members of the Organizing Committee contrived the appearance of such a condition.

Nobelist Baltimore launched the affair by announcing that ethical considerations would not be considered, nor would the question of whether hybrid DNA molecule experiments should be banned altogether. There would be no transcript of the proceedings, he said, but a tape recording would be made for archival purposes only and anyone who didn't want his remarks taped need only say so and the tape recorder would be turned off.

The tape recorders of the newsmen were already going, in plain sight. An obviously dismayed conferee called out, "What about the press?" Lewis diplomatically intervened and a show of hands settled that question in favor of the press. But Baltimore wasn't satisfied. He later complained to Lewis: "This is terrible. Every time you talk to somebody there's a goddamn ear there. Too many reporters just ruin this thing." When Lewis asked him to define the problem more precisely, Baltimore said that scientists were accustomed to use vituperative language in their discussions, and reporters would misinterpret it. Lewis assured him that experienced newsmen were thoroughly versed in screening out meaningless profanity and invective.

There was, in fact, so little friction between the newsmen and the scientists—in spite of daily press briefings from 2:30 to 4:00 P.M.—that Lewis quickly obtained Berg's agreement to admit any qualified reporter who happened to show up. And five such did appear.

One was a woman from the Paris *L'Express,* who flew from France to California without an invitation because she thought such an important conference must be open.

While she was expressing her astonishment and indignation over the ususual coverage rules to anyone who would listen, Kevin Howe, a seven-year veteran on the reporting staff of the *Monterey Peninsula-Herald* happened by and listened. He had no idea who she was or where she came from except that she spoke with a marked French accent. But what she was saying surprised him too.

The *Peninsula-Herald,* a daily with a circulation of about 33,000, is, as its name indicates, published in the city of Monterey, which lies about three miles from Pacific Grove. Albert Cross, the paper's managing editor, had received a query from the Associated Press office in San Francisco. Was the *Peninsula-Herald* planning to cover that big genetics meeting at the Asilomar Conference Center? Cross hadn't known that any genetics meeting, large or small, was taking place. He asked Howe to look into it. Howe telephoned the Asilomar manager, Roma Philbrook, a popular character on the peninsula. Was a big genetics meeting going on out there? Yes, Philbrook said, one was. It had started that very morning. If the *Peninsula-Herald* was interested, why didn't Howe come out and visit?

Howe went, and immediately encountered a group of reporters clustered around the woman with the French accent, who was volubly seeking the photographer who had been assigned to be there to help her. She hoped that he might be able to explain the curiously cloistered atmosphere.

Howe had been a reporter long enough to know that there are times to ask questions and times not to ask questions. This moment, he decided, was one of those when you don't ask. He listened instead, and then looked around. Piles of papers were stacked on the steps of the chapel where Asilomar meetings usually were held. He took a paper from each pile and went to a quiet spot to read them. Very quickly he grasped that the people gathered there were worried about *E. coli.* He knew a little about *E. coli* because the city of Monterey had its share of water pollution problems, and the *E. coli* count was always prominently mentioned in the warnings issued by the local public health officer. As soon as Howe felt he had enough information to go on, he went back to the *Peninsula-Herald* office and wrote the piece.

The *Peninsula-Herald* belongs to a newspaper chain whose leading paper is the *Toledo Blade* in Ohio. Science writer for the chain is Michael Woods, who works out of the *Blade's* office in Washington, D.C. Howe Telexed his piece to Woods for criticism and approval. Woods corrected a few technical inaccuracies and sent the item back to Howe with an okay. The *Peninsula-Herald* printed the story that afternoon, and a copy went to AP in San Francisco and out from there on the AP wire. Some of the papers on the AP wire had writers under the required wraps at Asilomar. Editors of those papers were understandably annoyed by the AP exclusive. How come? Over the phone that night, the question burned the ears of the fellows covering the Asilomar proceedings.

AP's appetite had been whetted. It asked the *Peninsula-Herald* for anything more that might be available. Howe returned to Asilomar the next day and again found a clump of reporters with their heads together. "Who blew it?" they were asking each other suspiciously.

"Who blew what?," Howe asked himself. In time, he figured out that they were talking about the AP item that he had been the source of. What should he do now? He wandered among the trees, absorbing the beauty of the place, and came upon Howard Lewis, who explained the rules the press was working under at the DNA conference. "He told me he didn't like the idea of exclusivity," Howe remembers, "but there it was."

Howe stayed at Pacific Grove on that second day long enough to interview a bacteriologist from the University of California campus at Davis, Professor Mortimer P. Starr. Starr was not impressed by the recombinant DNA research guidelines that were being debated that day. He said nature was recombining DNA molecules all the time, and if the scientists were to try to protect against all those happenings, "everybody would have to go around wrapped in cellophane." Howe quoted him in another article in the *Peninsula-Herald* on the afternoon of the interview, and didn't return to the International Conference on DNA Molecules.

It was just as well that Howe did not go back again. Stanley Cohen's nerves were already on edge; they had been from the start of the conference. He ate Maalox tablets continuously to keep his stomach acid down. When he heard about Howe's first scoop he got so angry he shielded his face from photographers.

The excitement generated by Howe's enterprise was the only break in the placid surface of the Asilomar press coverage. The presence of another uninvited observer from the media of mass communication passed unnoticed by all except Lewis. A TV man with a minicam drove up to the Asilomar chapel steps in a small panel truck with a pipe sticking out its back door. He shot human interest scenes through the pipe until he ran out of film. Only when he sought a source of more film was his mission recognized.

The woman from *L'Express* finally got the story she wanted, and pictures to go with it. But another competent female reporter's interest had a less happy outcome. She represented the Canadian Broadcasting Company. Not knowing that the coverage limitations would be relaxed during the conference, she abided by the original restrictions and stayed home.

New Times, likewise, offered no challenge of its exclusion, even though its counterculture rival *Rolling Stone* got in.

When the conference ended on February 27 and a press wrap-up was held in San Francisco, Lewis counted himself and the academy lucky to escape an embarrassing dilemma. He didn't know—perhaps still doesn't—of one very serious miscarriage of common sense.

Berg had felt that the scientist who challenged the SV 40-lambda phage-*E. coli* experiment in June 1971—Dr. Robert Pollack, who in late 1974 was preparing to leave the Cold Spring Harbor Laboratory to accept an associate professorship on the Stony Brook campus of the State University of New York—should be recognized by an invitation to Asilomar. Pollack knew how deeply divided the scientific community was over the hybrid DNA molecules issue. He didn't feel that much of a decisive nature could come from Asilomar. He thought the greatest contribution anyone could make toward resolution of the hybrid DNA molecule controversy was to help maintain an intelligent dialogue between scientists and the press. He therefore suggested to Berg that the seat offered to Pollack be given instead to Horace Freeland Judson, a former *Time* editor whom Pollack had recently met. Berg agreed to approve the transfer of credentials if Baltimore would endorse it. But Baltimore refused to sanction the exchange because Judson had no academic degree in science.

With that arbitrary and capricious decision, *Harper's* magazine, one of America's most distinguished intellectual journals, was deprived of the opportunity to include a firsthand account of the Asilomar proceedings in a two-part series of articles it had contracted to buy from Judson. Those articles were published later in 1975. In one of them, Judson disclosed publicly for the first time that it was Pollack who lit the fuse leading to the Asilomar revolution in science's relationship to the people.

Apart from the mishandling of the press, the Organizing Committee of the conference was guilty of other communication bumbles. One was discouraging the attempts by Professor Harlyn O. Halvorson and Dr. James McCullough to obtain an invitation. Another was the denial of a request from Professor Roy Curtiss III that would have broadened the political territory of the meeting to include Eastern Europe and thereby directly inform scientists who otherwise could get their information only secondhand through the Russians, who had been invited. Curtiss's proposal would have given Czechoslovakia representation in the person of Dr. V. Krcmery, who two years later organized a gene-splicing symposium in Prague that brought together scientists from both sides of the Iron Curtain, including East and West Germany. The single-minded handlers of the Asilomar guest list might have had difficulty in grasping the implications of that one. But there could be no excuse for the treatment of Halvorson and McCullough. Halvorson was an officer of the American Society for Microbiology, an organization whose expertise had been rejected by the recombinant DNA research enthusiasts as far back as the initial meeting of the Berg committee. McCullough, a senior member of the staff of the Congressional Research Service of the Library of Congress, has followed developments in genetic engineering for years. He is a frequent source of advice to concerned committees of both the

Senate and House of Representatives. Keeping him away from Asilomar amounted to shutting the eyes of the Congress to what was happening behind the closed doors of science.

22

PROFESSOR ROGER DWORKIN was born in Cincinnati, Ohio, in 1943. He attended Princeton University, got a law degree at Stanford, qualified for the California bar, and entered private law practice in that state in 1966. Two years later he accepted appointment to the faculty of the University of Indiana Law School. He has remained on the Indiana campus at Bloomington since then, taking occasional leave to do research in the law as it applies to biology, paying special attention to the response of legal institutions to rapid changes in life-style occasioned by scientific discoveries and their technological aftermath. One such excursion during the academic year 1974–1975 took him to Seattle to look into an intriguing legal dilemma, using the Medical School of the University of Washington as a base of operations.

When he left Bloomington on that occasion, Dworkin knew nothing about gene splicing, having spent most of his research time on such more familiar problems as human experimentation, organ transplants, definition of brain death, amniocentisis and other interventions in control of hereditary disease, genetic counseling, and the like. But after his arrival in Seattle his interest in hybrid DNA molecules was tweaked by a University of Washington geneticist, Stanley Gartler. Gartler wasn't involved in recombinant DNA research himself, but, like virtually everyone else on the campus, he was acquainted with Professor Stanley Falkow's work with antibiotic resistant plasmids and the enterotoxin gene of the *E. coli* bacterium.

Early in January 1975, Dworkin accepted Gartler's introduction to Falkow and paid a call to the Falkow lab. For more than an hour the famous pursuer of the diarrhea gene tutored the young lawyer in the peculiar ways of *E. coli* and other microorganisms. As he prepared to leave, Dworkin asked Falkow what were the chances of getting into the February conference on recombinant DNA molecules at Asilomar.

"I don't think the chances are very good," Falkow told him. "Attendance is by invitation. Budgeting requires that attendees be limited to the neighborhood of one hundred and fifty and applicants are already being turned away in droves."

Noticing the crestfallen look on his visitor's face, Falkow said, "I'll be happy to give you Paul Berg's telephone number if you'd like to try him."

Dworkin took Berg's number and returned to his own office to make the phone call.

"Berg was cool at first," Dworkin recalls. "But I wanted very much to attend that conference, and I wrote him a letter the next day, enclosing my curriculum vitae along with the names of a couple of people who, I felt sure, would be willing to vouch for me."

Toward the end of January, Dworkin appeared in Falkow's lab again.

"I want to thank you for your kindness," he said. "I've been invited to Asilomar, and I've been asked to speak."

"You've been asked to speak?"

"Yes," said Dworkin, "I've been asked to speak. The best that Berg felt he could do was put me on a list of nonscientist alternates. But he referred my letter to Daniel Singer, a prominent Washington lawyer—the husband of Maxine Singer. Singer is arranging nonscientist participation at Asilomar for Berg. A couple of the participants will be members of the bar. And Singer wants me to speak on some aspects of the law that scientists seldom think about."

Falkow couldn't believe it.

But the next time he saw Dworkin was at Asilomar.

Only after the fact did either of the two men realize that in the most casual fashion they had set in motion a train of circumstances that would shape science's intercourse with society for years to come.

23

WE HAVE seen, in the statement that Dr. David Baltimore wrote for the press conference at the National Academy of Sciences in July 1974, that the Berg committee never intended to consider stopping experiments with hybrid DNA molecules. On the contrary, the committee's purpose was to justify the research to the American people—who would have to pay the bill for it—and so to forestall possible interference with the work in the Congress or in the courts. No thought was given to asking the people how they felt about it, the assumption being that the people didn't know enough to make a reasonable decision. The Berg committee strategy was to put together the best case that science could muster and then depend on popular faith in the superiority of scientific judgment.

The Berg letter, published after the July press conference, had asked the director of the National Institutes of Health to name an advisory committee (on the unstated supposition that the committee would consist of the concerned scientists) to establish guidelines under which hybrid DNA research would continue. The NIH director had no authority to do this. But it could be done by the head of the U.S. Department of Health, Education and Welfare (of which NIH is part), and HEW Secretary Caspar W. Weinberger chartered the Recombinant DNA Molecule Program Advisory Committee on October 7, 1974. But before the guidelines could be written, scientists involved in DNA molecule hybridization had to reach a consensus on the prospects and dangers of the new technology. To bring the scientists together, the International Conference on Recombinant DNA Molecules was summoned in the last days of February 1975.

Rigorous logic would have required the conference to begin its assessment with the fundamental question: Was the totality of available knowledge so sparse as to suggest restricting all experiments to supersafe laboratories—as had been done with nuclear research—until no serious question of public safety remained unanswered? Alabama University Professor Roy Curtiss had advocated that course, as we have seen, but he was a relative unknown in the world community of science. As a lone delegate to the conference, he had virtually no influence beforehand. Had he been joined by the group of scientists who call themselves Science for the People, the effect would have been quite different, for Science for the People believed in getting down to fundamentals. But Science for the People was not represented at the International Conference on Recombinant DNA Molecules.

It will be recalled that while the Berg committee was still taking shape, Dr. Richard Roblin argued that unconventional thought should be invited into the discussion of pros and cons of gene splicing. A provocative thinker who impressed him was Dr. Jonathan Beckwith, of the Harvard University faculty, one of the founders of Science for the People. Beckwith is unquestionably one of the most gifted living molecular biologists. Peers describe him as "fantastically skilled." A fellow founder of Science for the People, Dr. Ethan Signer stands in the minds of many colleagues very close to Beckwith in the hierarchy of imaginative biology.

Roblin's pleadings were ignored at the time he first made them, but in the last-minute flurry of reconsideration that granted Indiana law Professor Roger Dworkin's request for an invitation, Beckwith was belatedly invited too. Beckwith, however, refused to be an "Uncle Tom." In his opinion, there was no time for tokenism. He believed that the Asilomar conference should include a half-dozen specialists in public health, a half-dozen epidemiologists, a couple of labor unionists, some officials responsible for enforcement of the Occupational Safety and Health Act, and perhaps a social historian. One person from Science for the People could not pretend to compensate for the absence of all those others.

After Beckwith declined what he saw as an impotent involvement, Dr. Jonathan King, another prominent member of Science for the People, was invited to Asilomar in Beckwith's stead. King's invitation reached him only five days before the conference was due to begin. He had prior commitments. King therefore suggested that his invitation be transferred to Dirk Elseviers, a California member of Science for the People, whose home was conveniently near to Asilomar. Berg and Baltimore wouldn't accept Elseviers because he was only a postdoctoral student.

Among the scientists who were invited to the historic event at the Asilomar Conference Center near Monterey, California, some believed that their freedom to experiment as they pleased was a legal right. Others accepted their responsibility to society in principle but did not believe there was an immediate need for control of hybrid DNA work.

Undoubtedly the normal defensiveness of the scientists on the issue was intensified by the speed with which the British government had stepped in in response to the Berg committee's moratorium appeal by appointing a Parliamentary Working Party to investigate the situation surrounding gene splicing. There was serious cause for concern in Britain just then because of a smallpox virus escape that no one thought could happen. Four people died in that episode, which started in one of the safest laboratories in the United Kingdom. The Parliamentary Working Party recommended controls on hybrid DNA molecule experimentation even before the conference at Asilomar got under way.

Confronted with regulations of yet undetermined severity, British scientists naturally were eager to show that they were responsible citizens. As early as October 1974, the British journal *Nature* published reports of tests to which scientists in Great Britain had subjected themselves to demonstrate that fears of using the *E. coli* bacterium's lambda phage as a gene-splicing vehicle were exaggerated. The drinking of quantities of lambda produced no ill effects. "Controversy," the British Broadcasting Company's social conflict program, invited Berg to debate the moratorium with Dr. Ephriam Anderson, head of the laboratory where responsibility for control of intestinal disease in the United Kingdom is lodged, and there were those who thought that Anderson got the better of the argument.

On the American side of the Atlantic, some researchers were fearful that what had begun as a voluntary, temporary measure might be turned into a permanent imposition. On the eve of the Asilomar meeting, both Berg and Maxine Singer had grave doubts that a majority of the assembling scientists (90 from laboratories in the United States, 50 from 17 foreign nations, including Australia, Japan, and the Union of Soviet Socialist Republics) would support continuance of the research pause.

There was one possible escape from this embarrassing dilemma. That was to seek agreement on the continuance of restraint only as long as it would take to remove the cause of fear of possible hazards in gene splicing. It had been suggested that the cause could quickly be removed by biological containment, that is, by "disarming" the microorganisms that do the splicing, making life outside the laboratory impossible for them. We saw earlier that the Berg committee at its very first meeting had ranked biological containment second on the committee's list of objectives but had neglected to mention it in the letter voicing the limited voluntary moratorium call. We also saw that specific proposals for achieving biological containment were received by the committee from three different sources, yet the committee had failed to provide room for discussion of biological containment on the program of the Asilomar conference.

"Disarmament" won a place as a conference subject through a series of accidents, beginning with the appointment of Professor Curtiss to the Asilomar panel on plasmids, chaired by Dr. Richard Novick. Other members of that panel were Professors Cohen and Falkow (whose *E. coli* panel meanwhile had merged with the plasmid panel) and an unflappable British type, Professor Royston C. Clowes, a genius of compromise from the University of Texas at Dallas, whose presence proved invaluable. All these men also belonged to the international plasmid nomenclature panel named at the Japanese-American plasmid conference in Honolulu in 1972. They were accustomed to working closely together. Their common experience made them conscious of loose organization around them before others were aware of it. They assumed responsibility for bridging the gaps rather than courting failure of the Asilomar meeting, which they hoped would assuage a fear they shared: that if the accelerating spread of antibiotic resistant plasmids was not soon stopped, the era of antibiotic medicine might end precipitately with nothing in sight to substitute for the abused "wonder" drugs.

As early as November 1974, Novick summoned the plasmid panel to New York City and there set up a system of periodic communication with the other panel heads. The intermittent messages disclosed an uncomfortable situation. Berg had asked Novick to have the plasmid panel draft a set of guidelines to govern use of plasmids in recombinant DNA experiments. Novick had assumed that parallel assignments were given to the chairmen of the other two Asilomar panels, Dr. Aaron Shatkin of Rutgers (for viruses) and Dr. Donald Brown of the Carnegie Institute in Washington, D.C. (for eukaryote-prokaryote hybrids). But the information Novick was getting from Shatkin and Brown indicated no such understanding on their part. Had Berg meant the plasmid panel guidelines to cover all experiments with hybrid DNA molecules or only experiments with plasmids? Recognizing that three different sets of guidelines might be difficult to mesh into a consistent pattern, Novick decided to let his panel do the whole job.

The guidelines were written into a plasmid panel report, which began with a statement of twofold purpose: "first, to explore and detail the potential biohazards posed by a wide variety of classes of experiments involving recombinant microorganisms so as to raise the general level of awareness of these biohazards; and second, to make available suggestions for dealing with potential biohazards so that the individual need not rely entirely upon his or her own judgment."

The gene-splicing technology was described as a "major breakthrough in molecular biology," which, along with its great potential benefits, carried "at least the potential of serious and often unpredictable adverse consequences" and "these possibilities have given rise to a significant level of concern among the general public as well as within the scientific community. . . . There is ample precedent for the fear that the accidental introduction of organisms into new environments may have uncontrollable and sometimes dramatic untoward consequences. As examples, one might point to fire ants, killer bees, mudfish, snails, *Xenopus* toads and to Chestnut blight and Dutch elm disease. More germane, perhaps, to the present document is the serious biohazard inherent in the astonishing spread in the space of a mere 30 years of bacterial plasmids carrying resistance to antibiotics, consequent to the vast overuse and misuse of these valuable therapeutic agents."

A set of principles, "some of which are clearly established as facts, while others may be regarded as assumptions," was put forth:

1. Since man has some measure of control over his actions, there is an operational dichotomy between the activities of man and the processes of the natural world. The distinction between "man-made" and "natural" is therefore meaningful and control of the former is both worthwhile and possible.

2. It is possible to modify profoundly the genome of a microorganism by artificial means involving the *in vitro* joining [that is, in the test tube] of unrelated DNA segments.

3. Modified microorganisms may behave in an unpredictable manner with respect to the expression of foreign genes, and to the effect of this expression upon their ecological potential (including pathogenicity).

4. The genetic effects of these manipulations may be different from anything that ordinarily occurs during the natural process of evolution.

5. Historically unforeseen ecological effects of technological developments have been more often than not detrimental to man and his environment.

6. The release of a self-replicating entity into the environment will prove to be irreversible should that entity prove viable in the natural environment.

Finally came the proposed guidelines themselves. These, the report said, were "designed to help the investigator perform responsibly and with confidence those experiments deemed sufficiently important to justify whatever risk may be involved" but "are not intended as a license to do unrestricted experimentation."

Hybrid DNA molecule experiments were categorized in six classes. Specific examples of each class were given. The classes were graded upward depending on the degree of their suspected risk. Experiments were allowable under differing degrees of physical containment in the first five classes. In the sixth class the "biohazards are judged to be of such great potential severity as to preclude performance of the experiment at the present time under any circumstances, and regardless of containment conditions." Cited as an example of Class 6 was the "introduction by any means of the genes for *botulinum* toxin [food poisoning] biosynthesis into *E. coli.*"

Having already risked censure by putting such a broad interpretation on its formal assignment, the plasmid panel hesitated before going still further out. But out it did go, and in Section E of Appendix C, on page 31 of a 35-page report, the panel spelled out a few of the biological containment ideas that Curtiss had urged upon Falkow the previous October. "Suggestions for Possible Genetic Modification of Recipient Strains" was the heading of Section E of Appendix C, which listed nine possibilities, all modifications of the *E. coli* bacterium:

1. Use of a mutant incapable of making purine, a chemical constituent of the genetic base alphabet letters A and G. Such mutants of many pathogenic microorganisms are not virulent.

2. Use of a mutant incapable of making diaminopimelic acid, the principal chemical element in bacterial coats. Diaminopimelic acid is not very prevalent in natural environments. If escaping bacteria are unable to find it, they will die.

3. Use of a mutant that cannot grow at 37° C. Because 37° C is the

normal body temperature of animals, such mutants cannot live in animals.

4. Use of a mutant that cannot grow at temperatures below 32° C. It could not long survive in soil, water, and other natural environments.

5. Use of a strain that would be unable to ferment or utilize a diversity of carbohydrates, and so would die in many ecological habitats.

6. Use of a mutant sensitive to ultraviolet light. Such a mutant cannot survive exposure to sunlight.

7. Use of a mutant unable to receive genetic information from other organisms.

8. Use of a mutant that is sexually sterile, hence unable to transfer plasmids to other bacteria.

9. Use of a mutant that is resistant to a multitude of potential transducing phages and therefore immune to acquisition of DNA from, or transmission of DNA to, viruses.

When read carefully from beginning to end, the full report, though it was not a perfect example of English grammar or of precise scientific definition, was impressive. Its completion was planned on a time schedule that would allow copies to be distributed at the Asilomar registration desk as the invited scientists arrived during the daylight hours of Sunday, February 23, 1975. The last of six days of work on the writing and editing was done at Stanford on Saturday, February 22. Saturday night was set aside for a copying marathon. At four in the morning on Sunday the copying machine quit, and the job had to be finished by hand. Consequently, the copies didn't get to Asilomar until Sunday evening, far too late to be read, much less considered, by the jet-lagged researchers before the sessions opened at eight thirty on Monday morning.

Thirty-five pages of closely reasoned text, set forth in single-spaced typing, therefore lay waiting to be digested while the first day's speakers played the scientific version of "show and tell." *E. coli* and its phage, phage in general, restriction enzymes, and eukaryotic biology were covered at the speed of a TV newsbreak. Novick and Curtiss were bracketed on the late Monday afternoon program to talk about plasmids.

Novick, as the plasmid panel chairman, spoke first. He used up all of his allotted time and some of Curtiss's as well. Curtiss, from his political experience at Oak Ridge, had learned the importance of timing. Instinct warned him that chances were against most of the scientists present setting aside enough time to read through the delayed panel report. They wouldn't know what was on page 31 unless they got a pointed prod to look there. So, in the few minutes still left to him to discuss sexual transfer of plasmids by bacteria, Curtiss threw an extraneous slide on the lecture screen. The slide contained ten lines of type:

GENETIC MANIPULATION OF BACTERIAL
HOST TO REDUCE BIOHAZARDS

1. Use of strain unable to synthesize purines
2. Use of strain unable to synthesize diaminopimelic acid
3. Use of heat-sensitive mutant
4. Use of cold-sensitive mutant
5. Use of mutant defective in carbohydrate utilization
6. Use of UV-sensitive, drug-sensitive mutant
7. Use of conjugation-deficient mutant
8. Use of phage-resistant mutant

If the slide aroused the curiosity of the audience, Curtiss planned to answer questions and thus open up a discussion of "disarmament." However, his hearers apparently thought he had made a mistake and let it pass in silence. Curtiss could only continue elucidating his assigned subject. His offbeat stratagem had failed to counter the inertia of the delayed report.

When Tuesday morning brought the plasmid panel members onstage to lead a discussion of the proposed guidelines, the spotlight that Curtiss had been so patiently trying to focus on the "disarming of the bugs" was grabbed by Berg's old BBC antagonist, Dr. Ephraim Anderson. Anderson was determined to press the advantage he had gained on Monday in the testimony of two highly respected British researchers who had swallowed glassfuls of *E. coli*—some in milk and some without milk—for a period of days without finding still living K12 in their feces. The intent was to raise doubt about how much K12 could survive passage through the human gut and therefore how much damage K12 hybrids might do in the gut. Still pummeling chief moratorium architect Berg, this time by proxy, Anderson asked how many of the plasmid panel members knew anything about epidemic disease from personal experience. All of them did, and Cohen and Falkow had spent years working in hospitals. Why, then, Anderson wanted to know, had the panel done such a sloppy job in its definitions?

"For our purposes, pathogenicity and virulence are defined similarly as the 'capacity to cause disease,'" he read from the panel report. "This must rank as the greatest oversimplification of all time."

Novick said it was hardly surprising that some of the definitions should need revising. "This is, after all," he said, "rather a terse document."

Anderson scornfully flipped the 35-page packet of paper in the palm of one hand. "You could have fooled me," he scalded.

Falkow observed that the report had been put together in only six days, and it wasn't possible to dot all the i's and cross all the t's in six days. "Why couldn't you do it in six days?" Anderson demanded. "The Lord created the universe in only seven."

From the plasmid panel report definitions, Anderson shifted his attack to the report's grammar. "Conjugal" was clearly not the right word for bacterial sex.

The panel members were uncomfortable, and showed it. The spells of laughter provoked by Anderson's quips did the cause of Berg and Singer no good.

Nobelist Watson pulled down the curtain on Anderson's act by staging one of his own. He rose and addressed conference chairman Berg:

"I think we should lift the moratorium."

An almost visible shock wave ran through the hall. Watson was one of the eleven members of the original Berg commitee.

Maxine Singer jumped up. What, she asked, had happened in the prior six months to make the need for the moratorium any less than it had been when Watson voted for it and proposed the Asilomar conference?

In the experimental sense, nothing had happened. Watson mumbled something about the impossibility of legislating against stupid lab practices, then sat down.

Joshua Lederberg, the only other Nobel laureate at the conference, rose to fill the gap left by Watson. Lederberg objected to guidelines of any kind. They would quickly freeze into law, he predicted.

A fellow Californian, much younger than Lederberg, wasn't impressed by Lederberg's credentials. "He's not a bench scientist any more," the youngster told one of the newsmen. "I hope we're not about to hand a brick to some people who would like to have a brick to throw at science."

As of that moment, David Baltimore looked like a wise man. In introducing Berg as conference chairman at the opening session on Monday, Baltimore had said that the outcome of the conference would be determined by consensus. Asked what the criteria would be for determining the consensus, he replied that they would depend on the degree of the consensus. He was plainly setting the stage for avoidance of a vote if the vote seemed destined to be one of "no confidence" in Berg and the Organizing Committee. Almost everything that had happened at the conference since those opening remarks suggested that that kind of vote was coming.

In the gathering gloom one electrifying person stood out. The hall lit up whenever he spoke. Attending newsmen couldn't make up their minds whether to call him a gnome or a leprechaun. His bushy eyebrows, mobile face, and exaggerated body English fit either designation. He was the only foreign member of the Organizing Committee—Dr. Sydney Brenner, a gifted researcher as well as a witty one, with a superb command of the language. Brenner had preceded Curtiss in trying to interest the original Berg committee in biological containment. Even before that Brenner had tried to interest the British Parliament.

At the Asilomar session of Tuesday morning, Brenner made a profoundly moving plea for priority attention to means of "disarming the bugs." He flattered, wheedled, and cajoled those present into giving disarmament a place on the program. It was an offside place—just a workshop—and a cramped place, between lunch and the start of the 4:30 P.M. sessions on Tuesday and Wednesday. But it was a place. That one feat of persuasion alone deserved Stuart Auerbach's characterization of Brenner in the *Washington Post* as the "conscience" of Asilomar.

Forty to fifty scientists attended both sessions of the workshop. Prominent among them, aside from Brenner, were Clowse, Cohen, Curtiss, Falkow, Dr. Waclaw Szybalski of the McArdle Cancer Laboratory at the University of Wisconsin, and Professor Frank Young of the Rochester University Medical School. Means of disabling *E. coli* and *B. subtilis* bacteria and several species of phage were analyzed. Much of the talk was theoretical. But Curtiss described two existing cripples that he had created for other experimental purposes. One of these deliberately weakened strains of *E. coli* K12 lacked the gene that codes for the making of the bacterial coat; the other lacked the gene that controls secretion of restriction enzymes. Near the end of the Wednesday workshop, Curtiss offered to undertake a "disarming" mission with his Alabama laboratory's unspent NIH and NSF grant funds. The offer was accepted. Curtiss confirmed it with this longhand memo to Berg:

TO COMMITTEE ON RECOMBINANT DNA

Re: "Safe" *E. coli* hosts
I will make any of my present "safer" strains available immediately to any investigator desiring them. Work will be rapidly accelerated to construct and test strains in line with my report to you and to make each succeeding generation of safer strains available upon request. I

am hopeful that the accomplishment of this objective can be achieved within the next several months.

Roy Curtiss III

The International Conference on Recombinant DNA Molecules could now rightly claim to have initiated an exciting new biological concept. Would it be able to claim more? What about the guidelines?

The period just before dinner on Wednesday was a downbeat time. The mood was really a hangover from Tuesday, when the conference virus panel, whose area of responsibility covered DNA hybrids capable of entering animal cells, had rejected the counsel of Dr. Andrew Lewis, the National Institute of Allergy and Infectious Diseases (NIAID) research physician we met in chapter 2.

Lewis was the first scientist in America, if not in the world, who had tried to set up a system of voluntary control over DNA molecule hybrids created in the laboratory. Having been invited to Asilomar and appointed to the virus panel, he had felt "it would be useful to document and comment on the events, deliberations, and problems" he had encountered. He therefore drafted a report that began with the discovery in the late 1960s of a family of hybrid SV40 adenoviruses capable of infecting the upper respiratory tracts of children and also causing tumors in some small animals as well as transforming laboratory cultures of healthy human cells into cells that looked and acted like tumor cells. In September 1971 he distributed seed of those viruses to researchers at the Cold Spring Harbor Laboratory on Long Island and elsewhere along with letters specifying precautions it would be wise for experimenters to take and asking that seed of that seed not be distributed to other researchers.

When objections were raised to his procedure, Lewis decided that the problem should not be a personal problem of his but an administrative problem of the NIAID. He therefore wrote a memo in 1972 requesting a policy statement from the NIAID director. The memo passed through the hands of the NIAID scientific director, who referred it to several NIAID researchers and to members of the NIAID advisory committee, all of whom agreed that all experiments with the SV40-adenovirus hybrids should be done under laboratory conditions of strict containment.

Because Congress had given authority over interstate traffic in hazardous substances to HEW's Center for Disease Control in Atlanta, Georgia, rather than to the NIH, Lewis and a few other representatives of NIAID went to Atlanta to confer with CDC officials about measures appropriate for control of distribution of hybrid DNA molecules. CDC's method of dealing with hazardous substances in general was to issue permits to qualified handlers and levy fines of $10,000 or 10-year jail sentences on violators of the permits. That procedure applied to matters of immediate hazard. Lewis, however, couldn't demonstrate immediate hazard with hybrid DNA molecules. The hazard feared from them might not show up for ten to twenty years. So Lewis went back to Bethesda and drafted a "Memorandum of Understanding and Agreement." NIAID approved it in November 1973. NIH approved it in September 1974.

After telling the story up to that point to Asilomar conference chairman Berg, members of the Berg committee, and members of the Asilomar panel on viruses, Lewis went on to explain that the Memorandum of Understanding and Agreement attempted to restrict distribution and use of the hybrids to those laboratories that were willing to assume full moral and legal responsibility for containing the hybrids in the laboratory. Certain minimum containment standards were set.

"Several major laboratories . . . leaders in the field of adenovirus and SV40 genetics have, thus far, not supported the Memorandum of Understanding and Agreement mechanism," Lewis reported. This "in spite of the installation of adequate containment facilities and the presence of investigators in these laboratories who are quite interested" in the hybrids. To his knowledge, he said, "no explanations . . . or criticism of the wording . . . of the Memorandum" had been offered. Furthermore, he added, "one or more of these same laboratories . . . have supplied this virus [Ad2ND] to other investigators who may or may not share the concern about the risks."

"Significant implications for any attempts to deal with the problems posed by bacterial plasmid recombinants" could be seen in this experience, Lewis declared. Then he spelled them out:

> The ultimate containment of any infectious agent rests with the investigator who is studying the organism. Our experiences thus far indicate that it is unlikely in the competitive atmosphere in which science functions that broad unenforceable requests for voluntary restraint will contain the potentially hazardous replicating agents which arise from the widespread application of the plasmid recombinant technology. Since the creation and use of these recombinants may well represent the most hazardous technological advancement ever devised by biologists, I think most would agree that the initial applications of this technology will have to be carefully controlled and monitored. Rather than resorting to unenforceable voluntary mechanisms, perhaps influence can best be exerted at the level of individual institutions by project evaluation, peer review and moral and legal pressures. If not, it

> will certainly be necessary to consider nonvoluntary approaches based upon legally enforced regulations and licensing procedures.

Lewis emphasized that the opinions he expressed were his own and "should not be construed as reflecting the views of the administration of either the NIAID or the NIH."

The Asilomar virus panel, in its report to the conference on the night of Tuesday, February 25, disdained Lewis's advice. In essence, the panel declared that hybrid animal virus experiments could be safely conducted under the rules of research normally observed by the National Cancer Institute. In the only formal dissenting opinion to come out of Asilomar, Lewis wrote:

> I disagree. Given the limited amount of information available at this time, I believe that the risks associated with the widespread, semi-contained use of this procedure exceed the rewards from the information to be obtained. Because of these risks, the diversity of opinions among individual scientists that these risks have provoked, and the moral and legal issues raised by the rising public concern about rapid advances in biomedical technology, I believe that application of this procedure to study animal virus genomes requires a methodical and carefully conceived attempt to reduce the risks to more acceptable levels. The initial step in the reduction of these risks is the development and testing of new and theoretically safe vectors. After such vectors have been provided, laboratories with adequate containment facilities (equipped to handle moderate risk organisms as defined in the NCI "Guidelines for work with oncogenic viruses") could begin to study recombinants containing DNA from supposedly nonvirulent regions of the genome from low to moderate risk animal viruses while attempting to assess any hazards associated with such recombinants. As the problems associated with such agents become better understood, more detailed studies could be undertaken. By such a slow, step by step approach, I believe that useful data could be obtained, the risks posed by these recombinants could be adequately evaluated, and most importantly, any untoward problems could be quickly appreciated and contained.

The "disarmament" workshop on Tuesday had generated enough talk to arouse considerable sympathy within the conference for Lewis's view. On the conference floor, dissatisfaction with the position of the virus panel majority was vigorously expressed, and the panel members were exhorted to rethink their report. When the panel resumed its deliberations, however, Lewis still felt very much alone and believed his cause was lost.

Because of the way the press had been muzzled (willingly, it must be noted), there was no way for the outside world—the world that had to pay the bills and deal with untoward consequences—to exert any influence on the decisions that were about to be made inside the Asilomar meeting hall. For the public couldn't know until later what was happening inside the hall. The newsmen inside were committed not to report anything until after the conference adjourned. And after adjournment, the news would be too late to have any meaning.

That was how things seemed on the evening of Wednesday, February 26. But it wasn't quite the way things were. Four representatives of the outside world had been inside the hall from the start of the proceedings on Monday morning—four attorneys-at-law: Daniel Singer, Maxine's husband, and three other nonscientists whom Singer had persuaded Berg to accept. Four people whose minds were open to other methods of fact-finding as well as to the scientific method, which is not designed to consider qualitative but precisely unquantifiable facets of life such as people's feelings.

Singer had hoped that one of those four people would be Daniel Callahan, director of the Institute of Society, Ethics and the Life Sciences. But a group of North Carolina University students, who had no idea of the significance of what they were interfering with, refused to release Callahan from a commitment he had made to them almost a year before. With only a few weeks remaining before the conference opening, Singer had to find an appropriate substitute for Callahan. The choice was one of the institute's founders, Harold Green, a professor at the National Law Center of Georgetown University in Washington, D.C., a specialist in the analysis of risks versus benefits.

Green addressed the conferees on Monday afternoon. Because of the confusion produced by the delay of the primary document the conference was to act on, the timing backfired. An already restive audience drew a permissive inference from this set of Green's remarks:

> The present moratorium, and this conference, represent an admirable approach to the problem of protecting society against hazards inherent in scientific and technological advance. The risks of proceeding, including risks inherent in uncertainty, have been forcefully identified and emphasized. It is particularly admirable that this action has been taken by the scientific community itself rather than having the situation await more formal action by government. I do not know what the outcome of the conference will be. As a layman and, in a sense, an outside observer, I sense that the issues are complex and that there is no answer that is necessarily correct or incorrect. The important thing is

that the issues be fully ventilated, discussed, and debated, and if this is done, as I expect will be the case, I for one will feel comfortable whatever the resulting decision may be.

How little comfort Green really would have found in certain decisions the conference might have made can be judged from these other passages of his talk:

The problems under discussion at this conference are not unique. Comparable problems of balancing benefits against risks are found in many other areas in which science and technology are advancing. One element that is common to all of these areas is the fact that benefits are always relatively obvious, immediate, and intensely desired, while risks are usually relatively remote and speculative. There is, moreover, usually no constituency for the risks—very few people have the knowledge, resources, or incentive to press the risks upon decision-makers. Our major need is to find means through which risks are given time and dignity more equal to that given to benefits in the decision-making process.

* * *

If the objective is to prevent the devolopment of technologies that may have unacceptably adverse consequences, the optimum time, indeed perhaps the only effective time, to exercise control is at the inception of the development. I am not saying that control should be exercised merely because demonstrable risk, let alone purely hypothetical risk, exists. What I am saying is that there is a momentum inherent in science and technology—a momentum fueled by man's innate desire to improve his lot and by the ambitions of the practitioners of science and technology. This momentum has in the past led to commitments to technology that we have come to regard as involving unacceptable injury. If we desire to guide the development of technology so that we may enjoy its benefits free of unacceptable injury, we must find ways to intervene before momentum takes over. The enthusiasm and optimism of the proponents of the technology must be tempered at an early stage by a more deliberate, explicit, and somewhat more pessimistic consideration of the area of uncertainty as to potential hazards.

* * *

I recently had the opportunity to read the text of a paper presented by Freeman Dyson in Madrid last October. The title of the paper was *The Hidden Costs of Saying No.* Dyson argues in this paper that the costs of saying "no" with respect to scientific and technological developments may be greater than the costs of saying "yes" because, as he puts it, "the

> prohibition of dangerous experiments may well imply the postponement of an industrial technology that is of special importance to mankind." He suggests that scientific and technological advance is the product of a "yes" or "no" decision by some person or entity empowered to permit or ban the advance. This is in reality rarely the case. Scientific advance generally occurs because a scientist wants to do something and does it without any necessity for his obtaining a "yes" from an authoritative source except if he requests funding support from the government or some other external body or, as when human experimentation is involved, from an institutional review body. Except in such cases, moreover, there generally is no one to say "no" at the outset, and "no" will be said only at a much later time as society struggles, often with futility, in the effort to bring adverse consequences under effective control after significant injury has already occurred.

Treating Monday afternoon's negative response to Green as though it had never registered, Singer on Wednesday night announced his intention to "sharpen the focus of Harold Green's remarks" in order to "give you a framework for the deliberations of tomorrow morning." "Hopefully," he said, he would "suggest a broader perspective that would go beyond immediate realities to a future not yet clearly visible."

By instinct, Singer said, he considered it "wrong to prohibit intellectual activity," assuming that there was a public interest in that activity. "The public is paying for what you are doing," he went on. "How much, if anything, is it willing to pay?"

Singer exhorted his hearers to remember that they were "acting for others" who would want to know "what are the circumstances under which you can subject another human to non-trivial risk? What responsibility do you assume toward the victim of whatever injury may occur? How likely is the event to occur, in a cumulative as well as in an individual sense? How widespread is the opportunity? Is it controllable, reversible, compensible? Has it been analyzed and quantified as much as possible?"

It is not reasonable for any scientist to assume that any person who works for him "is willing to join you in martyrdom," Singer declared. "He must be capable of making a rational choice for himself. He must be able to recognize the possibility of an untoward event. You are answerable to him for all the consequences of your acts. You have an obligation to explain. You are responsible for his training, and for maintaining standards of good behavior for him, particularly when others are exposed. Ideally at least, ethical codes should govern. Your moral and sometimes your legal responsibility is similar to what it would be if you were keeping a lion instead of a domesticated cat as a house pet."

The learned professions have a special responsibility, Singer pointed out. "In standing before the law, an experimenter must be able to claim a right to expose others. He must accept some kind of procedure to legitimize conscious exposure of others to the possibility of harm." Two principles were to be remembered above all else:

1. "Proponents of experiments bear the burden of demonstrating the innocence of the experimenter. It is not the burden of the rest of society to prove the existence of danger."
2. "Big benefits do not justify big risks."

"I recognize that what I have said is not soothing to you," Singer concluded. "To some of you, it is probably distasteful and even hostile. I can only hope that my words will be received with good-will."

Singer then turned the lectern over to Alex Capron, professor of international law at the University of Pennsylvania. Capron immediately challenged Baltimore's opening emphasis on the scientific and technical aspects of the conference. "Why was so much time spent away from the task at hand?" he asked and answered his own question: "You did what you were accustomed to doing because it was easier than the real problem, which you can't solve by yourselves. You must share it with your government, with the public, with secretaries, technicians, and janitors."

The bombardment continued relentlessly. "Academic freedom is limited. Freedom of thought does not extend to the causing of harm." "This group is not competent to judge the risk-benefit ratio of experiments. That is a social decision, as is the judgment on benefit itself." "Great scientific interest is not enough." "Progress is only an optional value." "You must recognize the right of the public, through the legislature, to make what in your opinion are—and in fact may be—erroneous decisions. You have the right to lobby to influence those decisions, but you also have the obligation to abide by the decisions in the meanwhile."

Taking up the question of the guidelines that were to be considered the next morning, Capron said that the scientists who would make up the NIH Advisory Committee on Recombinant DNA Molecules were "not competent to handle this issue. No guidelines ought to be issued prior to their promulgation by a larger group. The vehicle of public advisement should not be the *Federal Register,* which is read by only a handful of people. You must use the committees of Congress, which in turn can use the Office of Technology Assessment or a national study commission appointed for the purpose."

One appropriate form of legislation could provide an insurance fund that would insulate individual scientists from financial responsibility, Capron suggested. Whatever laws were adopted, he said, should be uniform, preferably administered by the scientific community, with review by laymen and appeal to higher authority.

On the international level, Capron advocated parallel laws for different nations rather than a single overall piece of legislation. He suggested that the International Association of Microbial Societies might be a facilitating agency. Actual regulation should be left to individual countries.

As Capron sat down, Roger Dworkin, the Indiana law professor who had asked for the opportunity and had been put on a standby list by Berg until Singer intervened, rose to continue the message. "I propose to lower the tone," Dworkin said, and promptly did by describing "conventional tools of the law and how they may be used against you—in the form, say, of a multimillion dollar lawsuit."

The temperature in the hall cooled noticeably.

"Despite many innovations and changes," Dworkin said, as though he were lecturing a class of college students, the "tort system remains the major injury compensation law of the United States." It pays considerable attention to conduct. Conduct is not so difficult to modify if you have to pay for failing to modify it.

Two kinds of negligence are applicable to work with recombinant DNA molecules. One is negligence in performance. Was something done in the wrong way? Was a safeguard lacking? Then there is the second kind of negligence, which resides in doing anything at all. Either kind of negligence entails unreasonable conduct. This is not quantifiable. How likely is the conduct to hurt somebody? How bad is the injury inflicted by the conduct? How valuable is the work to which the conduct contributed? How expensive would it have been to do something else? A jury of laymen, expert in nothing, has a substantial role in deciding responsibility. The judge who will charge the jury will almost certainly not be a scientist. The judge may allow the jury to decide whether a questioned experiment should have been done at all.

There is a limit to responsibility. The law recognized that in the doctrine of the stopping point. It might help a gene-splicing scientist in a bizarre case: "Let's say if a burglar strewed your stuff all over Brooklyn. Maybe that would get you off the hook."

But something isn't necessarily safe just because it is reasonable. The work must be appropriate to the place where it is done. Indiscriminate blasting in downtown San Francisco isn't appropriate. What place is appropriate for recombinant DNA research? Universities in the midst of crowded cities? Or the wilds of Montana? We don't know.

The value of the research to the community can be a defense if anyone sues. To achieve that status, however, researchers must be more specific than they have been in documenting claims of the benefits of the hybrid DNA molecule experiments.

Claimants for workmen's compensation need only show injury on the job. Awards are virtually automatic. Amounts are usually small. But allowance must be made for inflation. The costs are assessed on the employer, not the employee.

Under the Occupational Safety and Health Act, the workplace must be free of recognized hazards that are likely to cause death or injury. *Free,* the law says, not reasonably free. *Recognized* may be interpreted to mean *should be recognized.* The law is administered by the U.S. Secretary of Labor. It calls for use of the latest scientific data and the highest feasible degree of safety. Record keeping is mandatory. Inspections can be made without notice. Penalties are both civil and criminal. One violation can bring a $10,000 fine and 6 months in jail. Do it twice and you get $20,000 and a year in jail. There are no exemptions either for universities or for science.

The question-and-answer session that followed Dworkin's talk left no doubt that he had made the deepest impression of the evening. The vision he had conjured of inspectors from the U.S. Department of Labor swooping down on research laboratories without warning and slapping fines or jail sentences on slovenly experimenters was just too much to contemplate with serenity, so alien was it to the permissive regulations of the NIH, which are promulgated at least as much by the researchers themselves as by the officials who are ultimately responsible. Dworkin said afterward that he could feel the resentment of the audience, which seemed to think he was exulting over the prospect of scientists getting a deserved comeuppance. All he was really trying to do, he explained, was to perform a good lawyer's function of laying unpleasant facts on the line. His personal attitude, he said, had been stated in his talk—that the law has a tradition of allowing the professions to govern themselves and that people will respect that tradition, but there is also precedent for disaster when the allowed self-discipline is not applied, as it has not been in the medical profession, where physicians are now "being massacred in the courts" for their failure to crack down on malpractice in their ranks.

The chastening effect of the lawyers' participation in the Asilomar deliberations was felt before Wednesday night was gone. The first benefit accrued to the ethical position taken by Dr. Andrew Lewis. The virus panel on which he sat reversed itself and adopted his minority opinion. Someone who tried to compliment Lewis afterward for his apparent persistence and persuasiveness was surprised to have him say that he didn't have much clout "but the lawyers had a tremendous influence."

Even the Organizing Committee didn't realize how great that effect had been until the next morning. The committee sat up half the night drafting a Provisional Statement of the Conference Proceedings. When Berg presented the five single-spaced pages of it at 8:30 A.M. he gave the conferees half an hour to read it in silence. The committee would take full responsibility for it, he said; it would not be a statement from the conference. Although he did not say so then, the fact was that he and the other committee members feared that if the statement were put to a vote it would be voted down.

But Professor Cohen, pointing out that no matter how it was presented the statement was going to be accepted by the public as an act of the conference, insisted on a vote. There was a show of hands for each of the six sections of the statement. Section 1 had something in it for everybody:

> The participants at the meeting agreed that the pause in research, called for in the July 14 committee letter, ought not to be left unresolved. . . . Ignorance . . . has compelled us to conclude that it would be wise to exercise the utmost caution. Nevertheless, the work should proceed . . . with appropriate safeguards. Although future experience may dispel many fears, standards of protection should be set high at the beginning and each escalation, however small, should be carefully assessed.

Anderson complained about grammar again, and wouldn't accept Berg's explanation that the committee was tired when the statement was written. The vote of approval was unanimous except for a lone conferee who felt that that opening section should prohibit some obviously very dangerous experiments under any conditions.

Berg was flustered by his own inability to judge the prospects, which he later blamed on the disproportionate attention he had paid to the supposedly superior voices of Nobelists Lederberg and Watson. Maxine Singer confessed that she had shown herself to be a "rotten politician."

Section 2 of the statement said that "though our assessments of the risks involved with each of the various lines of research on recombinant DNA molecules may differ, few, if any, believe that this methodology is free of any risk. Reasonable principles for dealing with these risks are to adopt containment as a part of the experimental strategy . . . and match the risk [with] the effectiveness of the containment. . . . Consequently, we must seek means for estimating the risk, perhaps subjectively at first but objectively as we acquire additional knowledge."

Factors defining containment were listed in this section: Education and training of all laboratory personnel. Suitable protective devices in the laboratory. Good microbiological practices. Lastly and most significantly, two types of biological barriers: bacteria that are "disarmed" for life outside the laboratory and plasmids and phage that have been made dependent on such bacteria for survival.

That some experiments are too dangerous to be undertaken at all was argued again in the voting on Section 2. Dr. Robert G. Martin, of NIAID, insisted that "there are some experiments that should not be carried out under presently available techniques."

But Section 2 was approved, as it stood.

Section 3 scrapped the six explicitly defined classes of experiments proposed by the plasmid panel in favor of four sets of "broadly conceived . . . parameters . . . meant to provide provisional guidelines." Low-risk experiments—those in which DNA was exchanged between simple organisms that naturally make such exchanges—could be done in ordinary microbiological labs. Safer conditions would be required for experiments that make organisms more resistant to drugs or increase the ability to harm man. Moderate-risk experiments—most hybrid DNA molecule experiments—could be done only under lab conditions now specified for the handling of viruses known to cause cancer in man. Hooded and vented safety cabinets would be required, negative air pressure would be maintained to prevent escape of organisms, and access to the lab would be limited. High-risk experiments, involving viruses that cause various diseases in man, could be done only under high containment, available in perhaps a score of places in the United States.

Although Lederberg complained ("very complex issues are being railroaded through [and] many points are not being heard and discussed"), Section 3 was also approved by a clear majority of the conferees, as were Sections 4, 5, and 6.

Section 4 dealt with implementation of the conference recommendations. A newsletter of some kind would be desirable to keep researchers abreast of developments in biological containment. Annual workshops would provide continuing education. Training of laboratory staffs would have to be looked to, emergency procedures would have to be taught, surveillance of the health of the personnel would have to be maintained, and high priority should be assigned to all "disarmament" work.

Section 5 proposed a "model containment review process." Because of its importance in any self-regulation system, it merits quotation in full:

> A review process should be established which would be able to determine whether a given laboratory has the appropriate containment facilities for a given type of experiment. As far as possible, the biohazard review process should not lengthen the time required for review of research proposals. The specific form of the review procedure in different countries for different scientific and industrial laboratories must depend on local circumstances. The following proposal is, therefore, presented as a model. The model is designed for universities in the United States but would have to be modified for other situations.
>
> Each university or research institution should have a committee empowered and trained to grade the physical containment facilities of its laboratories [e.g., low, moderate, or high according to established guidelines]. The local committee would provide the laboratory head with a statement certifying the containment rating of the laboratory [subject to periodic reevaluation].
>
> When an individual applies to an agency for funds to support work on recombinant DNA molecules, the certificate of containment rating would be appended. The group reviewing the grant would then determine whether the certified level of containment matched whatever biohazard might result from the proposed work. The biological barriers incorporated in the experiment, the magnitude of proposed growth of bacteria, the type of DNA to be cloned, etc., would all enter into the decision. If the reviewing group is satisfied, the grant would be processed for scientific merit in the usual fashion. If a question arises concerning the appropriateness of the certified containment level, the NIH Advisory Committee on Recombinant DNA Molecules or some other body would be asked for an opinion or ruling.

Section 6 listed some questions about gene splicing that remained to be answered. Can eukaryotic genes be *expressed* in prokaryotic hosts? Can free DNA molecules infect animals or plants? Can prokaryote-eukaryote recombinant DNA molecules, either free or encapsulated in phage, infect animal or plant cells and be expressed there? Can mammalian cells in culture be genetically transformed by free DNA? Can hybrid animal virus DNA or virus-plasmid hybrids cause tumors in animals? Can methods be developed to effectively monitor the escape of DNA hybrids?

The call for a final vote on the whole Provisional Statement of the Conference Proceedings provoked a long, hot debate over the repeatedly advanced and rejected proposal for a flat ban on research with organisms known to cause fatal and untreatable disease. Berg lined up with those who opposed such a ban on the grounds that it would violate freedom of scientific expression. But gadfly Anderson took the opposite side, saying that although he disagreed with the Provisional Statement, in many other respects he considered prohibition of obviously dangerous experiments imperative. In the end, the conference, with only five dissenting votes, instructed the Organizing Committee to insert the ban in the finished version of the document.

Cohen, Lederberg, and Watson were alone in opposing approval of the statement as a whole. Cohen insisted that he be recorded as voting against the signing of a document he had no opportunity to read.

Watson, who had proposed the Asilomar conference in the first place, afterward called the meeting a "waste of time." He said the assembled scientists "pretended to act responsibly but actually were irresponsible in approving recommendations that did not adversely affect the work of anyone present."

What Watson said was true. The Provisional Statement of the Conference Proceedings did not require any gene-splicing scientist to abandon any experiment that was under way. Yet Watson's indictment did not represent the whole truth. The whole truth had two more parts.

The first part was that the decision to flatly proscribe some experiments (those involving DNAs of highly pathogenic organisms, those involving toxin genes, and those calling for more than ten liters of cultures of hybrid DNAs capable of producing products harmful to man, to other animals, or to the plants on which all animals, including ourselves, depend) effectively breached the long established dogma that no source of knowledge should ever be closed to scientific exploration for any reason.

The second missing part of the whole truth about the Asilomar conference did not become apparent until June 6, 1975, when *Science* published a summary statement of a 250-page report that the Assembly of Life Sciences of the National Academy of Sciences had accepted from the Berg committee on the previous May 20. The summary included not only the ban on experiments clearly dangerous to man but rephrased the questions with which the Provisional Statement of the Conference Proceedings had concluded. The questions were finally worded this way:

> Nothing is known about the potential infectivity in higher organisms of phages or bacteria containing segments of eukaryotic DNA, and very little is known about the infectivity of the DNA molecules themselves. . . .
>
> Work should also be undertaken which would enable us to monitor the escape or dissemination of cloning vehicles and their hosts. . . .
>
> Little is known about the survival of laboratory strains of bacteria

> and bacteriophages in different ecological niches of the outside world. Even less is known about whether recombinant DNA molecules will enhance or depress the survival of their vectors and hosts in nature.
>
> These questions are fundamental to the testing of any new organisms that may be constructed. Research in this area . . . should be given high priority. In general, however, molecular biologists who may construct DNA recombinant molecules do not undertake these experiments and it will be necessary to facilitate collaborative research between them and groups skilled in the study of bacterial infection or ecological microbiology.

What an ironic commentary on the exclusion from the conference planning of Science for the People, which had pleaded in vain for a broad interdisciplinary approach to hybrid DNA research control!

24

IN 1970, the year before Dr. Robert Pollack's historic telephone call to Professor Paul Berg from the Cold Spring Harbor Laboratory on Long Island, another Long Islander—Dr. William J. Gartland, son of an old-fashioned family doctor—went to work for the National Institute of General Medical Services at Bethesda, Maryland. Gartland had a B.A. degree in chemistry from Holy Cross, a Ph.D. in biomedical science from Princeton, two years of postdoctoral training at New York University Medical School (where Pollack also had studied), and an additional year at the University of California, San Diego. He entered NIGMS as a junior administrator, processing grants and contracts for research in genetics. He was, then, in the right place at the right time for involvement in the hybrid DNA molecule research moratorium call issued by the ad hoc Recombinant DNA Molecule Advisory Committee of the National Academy of Sciences.

In July 1974, the director of the National Institutes of Health received the so-called Berg committee's request for appointment of a scientific panel to help write guidelines under which appropriately contained gene-splicing experiments could be continued without undue risk to the public health. The panel would require an executive secretary to keep loose ends from raveling. No one could then foresee how often or for how long the panel's advice would be needed. Gartland's duties at NIGMS were such that his time was extremely flexible, so he was picked to attend the September 1974 meeting of the Berg committee and participate in its deliberations.

On the following October 6, legal sanction for the advisory panel and for Gartland's position in relation to it was provided through the chartering, within the U.S. Department of Health, Education and Welfare (of which NIH is part), of a Recombinant DNA Molecule Program Advisory Committee for the purpose of supplying specialized counsel to the HEW secretary, his assistant secretary for health, and the director of the National Institutes of Health in three areas: "the evaluation of potential biological and ecological hazards of DNA recombinants of various types" in this country, the development of "procedures which will minimize the spread of such molecules within human and other populations," and the development of "guidelines to be followed by investigators working with potentially hazardous recombinants."

The charter said the committee would consist of a chairman and eleven others, to be selected by the secretary or his designees to serve overlapping four-year terms. Those members who were not full-time federal employees would be paid $100 plus per diem and travel expenses for each day of work. There would be approximately four meetings a year, all open to the public, all in charge of a government official, and all following agendas approved in advance by a government official. Records would be kept of the meetings. Management and staff services would be provided by the Division of Research Grants, Office of the Associate NIH Director for Scientific Review, who would designate the executive secretary.

Having already been chosen as executive secretary, Gartland could only wait for the appointment of the committee members (which occurred before the end of 1974) and accompany them * and the committee chairman, Dr. DeWitt (Hans) Stetten, Deputy NIH Director for Science, to the Asilomar conference. During the three and a half days of debate at Asilomar, Gartland had nothing to say, Stetten made only one brief statement. Whatever summary of the proceedings was to be issued, he said, the phrasing of it should be in "general terms. Specifics should be avoided if at all possible. The fewer regulations we have to live by, the better off we are."

* The original roster of the Recombinant DNA Molecule Program Advisory Committee of NIH included Dr. Edward A. Adelberg, Yale University; Dr. Ernest H. Y. Chu, University of Michigan; Dr. Roy Curtiss III, University of Alabama; James E. Darnell, M.D., Rockefeller University; Dr. Stanley Falkow, University of Washington; Dr. Donald R. Helinski, University of California, San Diego; Dr. David S. Hogness, Stanford University; John W. Littlefield, M.D., Johns Hopkins Hospital; Dr. Jane K. Setlow, Brookhaven National Laboratory; Dr. Waclaw Szybalski, University of Wisconsin; Dr. Charles A. Thomas, Jr., Harvard Medical School. All but Darnell, Littlefield, and Thomas had attended the Asilomar conference. Added to the roster later were Dr. Peter R. Day, Jr., Connecticut Agricultural Experiment Station; Dr. Elizabeth M. Kutter, Evergreen State College; Dr. Emmette S. Redford, University of Texas at Austin; Wallace P. Rowe, M.D., National Institute of Allergy and Infectious Diseases; Dr. John Spizizen, Scripps Clinic and Research Foundation; and Dr. Leroy Walters, Georgetown University. Falkow resigned at the end of 1976. Chu and Thomas finished terms of office in 1976.

Stetten spoke as a loyal adherent to NIH policy, which is oriented against regulation. The tradition is so strong that another official of the agency could later tell a recombinant DNA symposium without apologizing that NIH approached regulation as "porcupines approach lovemaking–cautiously." Stetten, however, had more than policy in mind. He had an ideal model for the governance of detail. It was the "Report of the Working Party on the Experimental Manipulation of the Genetic Composition of Micro-Organisms," prepared by a group of eminent British scientists appointed by the Advisory Board to the Research Councils of the United Kingdom in reaction to the Berg letter as it appeared in *Nature.* Headed by Lord Ashby, the Working Party had presented its findings to Parliament only a few weeks before the Asilomar conference opened. A classic example of pungent use of the English language, that document set forth the following seven recommendations:

1. Subject to rigorous safeguards .... these [gene-splicing] techniques should continue to be used because of the great benefits to which they may lead.

2. All those who work with these techniques should have training in the handling of pathogens and ... laboratories where the work is done should have the basic equipment for the containment of ordinary pathogens.

3. In laboratories using these techniques someone should be designated as a biological safety officer and ... some sort of central advisory service should be established.

4. Continued epidemiological monitoring of workers in the laboratories where these techniques are used should become a routine practice.

5. Large-scale work with any potential hazard involving quantities of 10 liters or more should not be done except in especially equipped laboratories.

6. Workers using these techniques should seek to reinforce the safeguards of traditional containment procedures by genetical devices to "disarm" the organisms they use for experiments and by conducting other kinds of research on possible hazards of the techniques.

7. Further investigations should be sponsored forthwith to improve our knowledge of the viability in the human intestine of laboratory strains of *E. coli,* especially K12, and their capacity to transfer plasmids to resident strains.

The Provisional Statement of the Proceedings of the Asilomar conference fell far short of either the brevity or the clarity of its British antecedent. Stetten hoped the gap would be closed in the guidelines his Recombinant DNA Molecule Program Advisory

Committee was to write. "Guidelines should be guides," was the way he saw it. "They should not be a dictionary, an encyclopedia, or a textbook."

The most conspicuous aspect of the initial meeting of the Program Advisory Committee, held in a posh San Francisco hotel salon on the day immediately following the close of the Asilomar conference, was Berg's absence. He had declined a place on the NIH committee because he wanted to continue experimenting with hybrid DNA molecules, and his participation in official decision making on guidelines for such research would in his view represent a clear conflict of interest. All the other members of the Berg committee had taken a similar stance, except for Stanford Professor Hogness, an intimate of Berg's, who had pioneered attempts at gene splicing a dozen years before Berg became interested in the possibilities. Hogness felt no qualms about the obvious conflict in his activities, but many other people did, and his presence on the committee fed rather than damped the internal disharmony that dominated the proceedings from the start.

Stetten's foremost problem in presiding as chairman was, in fact, the unpleasant reality that half of the committee wanted to do what he wished them to avoid—write a "dictionary, an encyclopedia, or a textbook." The other half of the committee saw no necessity for writing anything at all. The committee as a whole had only one common tie—a prima-donna style. "Everybody talked," Stetten reported afterward, "nobody listened."

After wrangling from nine in the morning to five at night, the committee agreed on only a few points of procedure: (1) hybrid DNA molecule experimenters seeking NIH funds in support of their work must first obtain the approval of an institutional biohazard committee responsible only for reviewing the safety of the workplace and the procedures employed there; the scientific merit of proposed experiments would be judged in the usual way by the NIH study sections, of which there are fifty, each categorized by a scientific discipline; (2) priority attention should be given to work on "disarming the bugs"; and (3) the Program Advisory Committee should be expanded to include an animal virologist, a plant pathologist, an epidemiologist, and a representative of the lay public.

Having accepted responsibility for encouraging recombinant DNA studies in ways that would not risk the public health, NIH did not dare to take the political gamble inherent in allowing any time to pass without some semblance of protective apparatus firmly in place. Stetten therefore pleaded that the Program Advisory Committee go on record with interim approval of the Asilomar guidelines. Arguing that no finished document was yet available for examination, the committee refused its endorsement except "in general." When Stetten went home that night he told his wife he despaired that the committee would ever agree on anything.

Stetten is a scholarly person, and a dedicated one, but his eyesight is weak and his reading speed is consequently far below the demands of intensive document study. If Gartland, working part time without a secretary of his own, was not to be hopelessly snowed under by communications from university laboratories eager to qualify for approval of upgraded experimental facilities, the process of guideline formulation would have to be expedited.

Through two more meetings, however, the members of the committee were if anything in angrier disagreement than before. At one point Falkow wrote to Stetten saying: "It would seem hypocritical to me to continue to serve on the committee" if efforts then being made to weaken the Asilomar guidelines should succeed. "It is my belief that our committee was, in part, an illustration of how scientists were capable of responsible administration of science-generated problems that could potentially have an adverse effect upon the community-at-large. I don't feel that these guidelines are a very good reflection of that concept. . . . I recall your own statement . . . that if something had any probability at all then it would in all likelihood occur, and that this should be a guiding principle of our deliberations. I don't feel the committee has followed that good advice." To Gartland, Falkow wrote: "I simply won't be a party to any effort to ease the way for a relatively small group of scientists to perform experiments that they feel are more important than any other consideration." In a later letter, Curtiss told Stetten that the amendments Falkow took umbrage at "are a license to experiment . . . in the almost total absence of controls and/or sanctions." Forty scientists attending a phage meeting at Cold Spring Harbor Laboratory agreed with those sentiments and recommended to Stetten that the NIH committee "should have much stronger representation from scientists not directly involved" in gene splicing and that there also should be "representatives of the public at large." Only after three successive drafts of guidelines were put together and torn apart did the committee finally reach agreement at its fourth session, in December 1975.

In a fantastic somersault that ended in one decision on one day and a diametrically opposite decision on the next day, the Asilomar guidelines were tightened instead of being loosened, especially insofar as they concerned so-called "shotgun" experiments with cold-blooded vertebrates such as frogs, fruit flies, and other insects. Hogness protested in vain. The committee further voted against permitting continuance of such experiments that had been started under the Asilomar rules. "Shotgun" experiments are those in which the entire genome, or total DNA, of an organism is chopped up and the pieces spliced into a bacterium one by one. Knowledge of what genes are in which DNA fragment is necessarily imperfect, hence the chance of unanticipated effects is greater.

Some press observers who attended that December committee meeting credited the achievement of new guidelines principally to the presence of Berg, Brenner, and Maxine Singer as observer-participants in the debate. Stetten, however, said afterward that the real catalyst was a two-day workshop that immediately preceded the gathering of the committee. At the workshop Curtiss reported success in his long struggle to design a strain of *E. coli* K12 so thoroughly enfeebled as to have virtually no statistical chance of surviving outside the laboratory. He called this "disarmed bug" χ1776—χ being the Greek letter khi or chi (short for Chicago, the university where Curtiss began serious study of molecular genetics) and 1776 being a two hundredth birthday salute for the year when the original thirteen American colonies of England declared their independence. Curtiss's hopeful assumption was that χ1776 would be officially approved for use during the calendar year 1976.

Those members of the NIH committee who felt that there was no real need to control gene splicing saw χ1776 as a means of escape from stricter regulation, while those who felt a continued need for safeguarding mechanisms saw χ1776 as a guarantee of safety. The existence of χ1776 therefore became the determining element in a system of containment that consisted of two parts, physical (P) and biological (EK, for *E. coli* K12). The new guidelines fixed four levels for P and three for EK, their use to depend on the real or hypothetical risks entailed in manipulation of different experimental organisms.

P1 required nothing more than standard microbiological practice. P2 required a few extra precautions, principally directed against dispersion of aerosols. P3 required negative air pressure to be maintained throughout the laboratory in order to prevent escape of any form of airborne life. P4 required air locks, protective clothing of special design, and showering by exiting personnel—the types of precaution taken by handlers of the most dangerous known pathogens. EK1 required the use of the standard K12 strain of the *E. coli* bacterium. EK2 required an *E. coli* K12 strain genetically altered in such a way that laboratory tests would affirm, on the average, survival of only 1 in 100 million escaping bacteria. EK3 required use of an EK2 whose disabilities were confirmed by feeding the bacteria to animals. EK2 and EK3 were not in reality single elements but systems in which not only the bacteria but the associated plasmids and phage were also crippled to prevent survival outside the laboratory.

Although unofficial, a table of the combinations of P and EK approved by the NIH guidelines brought no complaints from the interested parties following its publication in *Science* and in *Nature.* The entire package was opened to public view by the director of NIH at a meeting of his Advisory Council in February 1976, and he announced his approval on the following June 23. The research moratorium that the Berg committee had proposed and the Asilomar conference had endorsed was at last made official.

25

IN KEEPING with its long established policy of interfering as little as possible with the academic scientists to whom it grants public monies in support of their experiments, the National Institutes of Health under the original Asilomar guidelines for recombinant DNA research had approved plans for a P3 laboratory at Harvard University and had set aside upward of a quarter of a million dollars to pay for the project. Earlier approval had been given to the P3 lab designs of Paul Berg at Stanford and of David Jackson, Berg's postdoctoral collaborator in the now famous SV40-lambda phage DNA hybrid experiment of 1971–1972. Shortly after that experiment, Jackson had left Stanford for the University of Michigan to join microbiologist Robert Helling in creating the nucleus for a new recombinant DNA research center at Ann Arbor. Helling had been on sabbatical leave from Michigan's botany department at the time of his brief association with Herbert Boyer and Stanley Cohen in the hybrid DNA molecule constructions of 1972–1973. There appeared to be nothing unusual, then, about NIH's agreement with Harvard, which fitted the pattern that had been applied to Stanford and Michigan. The safety guarantees of the biohazard committee of each school had been accepted at face value.

What was the face value of the guarantees on Harvard's P3 lab? The question was raised on June 8, 1976, on the front page of the *Boston Phoenix,* a weekly newspaper. Charles Gottlieb and Ross Jerome, free-lance writers living in the Boston area (which includes the city of Cambridge, just across the Charles River from Boston), had written a thoughtful news report that ran across six pages under the headline, "Biohazards at Harvard." The subhead read: "Scientists will create new life forms—but how safe will they be?"

The NIH guidelines that had been debated at a public hearing in Washington, D.C., in the preceding February were due to be published during that month of June. Those rules required P3

James Watson *(left)*, and Sidney Brenner, who was on the Organizing Committee for Asilomar

Left to right: Norton Zinder *(standing)*, Paul Berg, David Baltimore, Sidney Brenner, Richard Novick, Richard Roblin, and Maxine Singer

DeWitt Stetten and Maxine Singer

Daniel Singer

Sidney Brenner and Roy Curtiss, III

Joshua Lederberg

Left to right: Howard J. Lewis, Director of NAS's Office of Information; David Perlman, reporter; Paul Berg

Photos by Andrew Stern. Courtesy National Academy of Sciences

SCENES FROM THE FORUM OF THE NATIONAL ACADEMY OF SCIENCES, MARCH 7–9, 1977: DEBATING "RESEARCH WITH RECOMBINANT DNA"

Erwin Chargoff

Stanley N. Cohen

Left to right: Robert L. Sinsheimer, Paul Berg, Alexander Rich

Herbert W. Boyer

Ruth Hubbard

George Wald

Jonathan Beckwith

Donald S. Fredrickson, Director,
National Institutes of Health

Photos by Paul Conklin. Courtesy National Academy of Sciences

A. Dale Kaiser *Courtesy National Academy of Sciences*

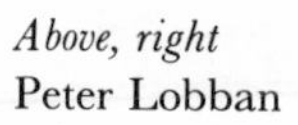

Above, right
Peter Lobban

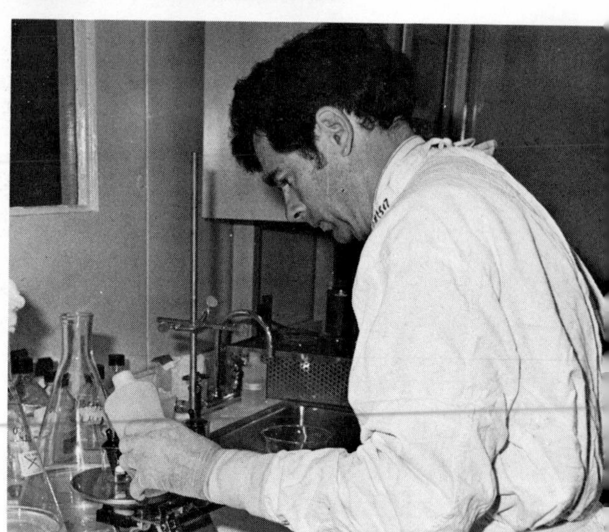

Andrew M. Lewis, Jr. *Courtesy National Institute of Allergy and Infectious Diseases*

Robert Pollack *Courtesy Cold Spring Harbor Laboratory*

containment for moderately hazardous experiments, for example, those involving genes of mammals other than man. In a P3 laboratory the two *Phoenix* writers reminded their readers, workers must wear protective gloves, must transfer their working materials in specially designed cabinets, must package and sterilize all experimental waste before disposing of it, and could remain at work only as long as the air in the lab was flowing negatively—that is, from the outside inward—to avoid escape of any microorganisms.

"Think of the worst possible place you could put this facility," Gottlieb and Jerome quoted Bill Petri, a Harvard Research Fellow, as saying. "You don't want to put it in an old building where the pipes break and cause floods. You don't want to put it in a place that is prone to electrical failures. You don't want to put it where there are large numbers of people going in and out."

Yet, the *Phoenix* report said, the Harvard Biological Laboratories building at 16 Divinity Avenue in Cambridge was an old building in which pipes often broke and caused flooding, in which electrical circuitry often failed, and in and out of which people were constantly moving. The fourth floor of that building was the site Harvard had chosen for the P3 lab.

Petri was not proposing to stop recombinant DNA experiments at Harvard. He believed such research "holds great promise." He was simply talking against putting the lab in what he thought was the worst possible place for it.

Gottlieb and Jerome next quoted Ken Miller, a Harvard biology lecturer who also favored recombinant DNA research. "They're planning the safest laboratory in the building," he declared. "But this isn't the safest building. If I came here at eight at night, I could open my desk and there'd be roaches running over everything. Heck, all the sugar in this building is covered so that ants can't get to it."

The ants were an apt example of the persistence of life in any ecological niche it finds suitable for its continued existence. They were not native to Cambridge. They were pharoah ants, the tiny *Monomorium pharaonis,* originally from Egypt, accidentally brought into the building at 16 Divinity Avenue many years before.

Regarding the indestructibility of the ants, the two *Phoenix* writers quoted from a letter that had been written to Daniel Branton, chairman of the Harvard Biohazards Committee, by Carroll Williams, Benjamin Bussey Professor of Biology, whose office was in the basement of 16 Divinity Avenue. Williams was a strong supporter of recombinant DNA research. He told his interviewers from the *Phoenix:* "No first-class department of biology or molecular biology can afford to pass this by. It's a dazzling opportunity." His letter to Branton concerned only the unsuitability of the Biological Laboratories building for the P3 facility, chiefly because of the ants:

"Exterminators recognize that this particular species is virtually impossible to eliminate by any presently known technique. If the protagonists of the P3 facility know how to keep the pharoah ant out of their area, I trust they will share their secret with the rest of us."

Branton, in turn, in a letter dated May 3, conveying the Biohazard Committee's approval of the siting of the P3 lab in the building at 16 Divinity Avenue, wrote to Harvard's Dean of Arts and Sciences, Henry Rosovsky:

"Our affirmation does not imply that we consider the fourth floor of the Biological Laboratories building the best or safest location for the laboratory. Nor have we addressed the question of whether there should be a P3 facility in the Biological Laboratories."

"Many of the inhabitants of the building" shared the view expressed by Williams, Miller, and Petri, the letter continued, adding: "Clearly a facility free of insects and not subject to flooding would be safer than one in which those risks must be contemplated. But, although there may be greater safety in having the containment laboratory in a different building, we have not evaluated the merits of any other location because no other specific location has been proposed."

Other Harvard officials who were questioned by Gottlieb and Jerome admitted that they knew of no way to eliminate the ants. The only hope they had was to limit the activity of the insects. A similar hope was all the university could offer in regard to the movement of people, who inevitably would carry some microorganisms out of the building on their hair and clothes.

Even Professor Mark Ptashne, who had applied for the NIH grant for the P3 lab on behalf of his friend, recombinant DNA researcher Tom Maniatis, agreed that "certain amounts [of the organisms] will be carried out on people's clothing." Ptashne argued that none of those organisms would be dangerous.

When the *Phoenix* writers wanted to know how the residents of Cambridge were to be protected from harm, Biohazard Committee Chairman Branton asked them:

"Why talk just about Cambridge? There are people from all over who come to Harvard every day. If an organism escapes, we might as well be talking about the world."

Professor Williams told Gottlieb and Jerome he might ask to have his office moved out of the Biological Laboratories building, even though the P3 facility would be four floors above him. The office of Professor Ruth Hubbard, wife of Nobel laureate George Wald, was directly on the P3 floor. But she said, "Picking myself up and leaving won't solve anything." Microorganisms that escaped from the P3 could be worse than the ants. If they once found a place where they could cling to life and breed, you'd never destroy them. And you couldn't predict how much farther than Egypt they might travel. Worse, you couldn't tell where to look for them because you couldn't see them.

The *Phoenix* report mentioned half a dozen internal Harvard meetings, held over a period of three years, to any one of which it would have been appropriate for Harvard to invite a representative of the people of Cambridge. No such invitations were issued. A small advertisement of one meeting had been placed in three local newspapers, but it stated the discussion subject in technical terms which, though understandable to scientists, did not suggest to the nonscientist majority that the public health might be involved. When, Gottlieb and Jerome asked Biohazard Committee Chairman Branton, would the Biohazard Committee address the neglected issue of public safety?

"I find it difficult to impose a level of hazard on society," Branton replied. "Somebody else has to decide that question. A more representative group has to be formed. The public must decide what level of hazard they find acceptable."

That kind of circular talk didn't make sense to Alfred Velucci, a man from an Italian neighborhood who had the gift of instant excitability. His personal theatrics, like the circuses of ancient Rome, were levers of political power; they had enabled him to remain mayor of the ethnically dominated industrial town of Cambridge (in the midst of which Harvard and Massachusetts Institute of Technology exist as islands of tax-free land) for twenty-six years. If Harvard's "guys in white coats," as he derisively called them, didn't know how to discover the feelings of the people of the city, Velucci did.

Beneath the mask of the loudmouthed absurdly gesticulating clown, however, was a shrewd and practical judge of situations. Some of those "guys in white coats" had formidable reputations. Among the ones he would have to tackle was a Nobel Prize winner. Other Nobelists around the country would rally around. Who of comparable prestige could Velucci count on?

While the mayor was mulling this question, he had a visitor, Nobelist George Wald, Professor Hubbard's husband. Wald looks like at least a bishop and maybe a cardinal. He wears his white mane long, so that it curls upward at the nape of his neck. A black frock coat hangs open to reveal a turtleneck sweater, likewise black, and above the sweater dangles a sedate gold medallion attached to a gold chain around his neck. The cathedral cadence of his speech once led someone to say, "When George talks to you, you have the feeling there is a crowd standing behind you."

Wald's purpose, it turned out, was to urge the mayor to take the report in the *Phoenix* seriously. If he did so, Velucci asked, would Wald support him openly? Wald said he would be glad to. When he walked out of City Hall, he left Velucci feeling "damn sure" of himself.

Immediately the mayor began to display the June 8 issue of the *Phoenix* in the City Hall offices as though it were a lantern from the steeple of the Old North Church across the Charles. He was alerting the countryside—or at least the excitement-starved news reporters—to the approach of "Frankenstein monsters" from Harvard Yard, which he once had threatened to pave over to provide much needed parking for motorists. (No one can now remember what reason he gave, on another occasion, for talking of turning the irreverent *Harvard Lampoon's* editorial quarters into a public urinal.) As befitted an emergency, real or imagined, Velucci next summoned an open meeting of the Cambridge City Council for the night of June 13. Whether through luck or inside information, he chose the night of the day on which NIH released the new recombinant DNA research guidelines in Washington. It was one of the grandest political coups—probably the very grandest—in Velucci's long and vividly complexioned career.

While the motley hundreds of citizens, who that night jammed the old wood-paneled City Hall to its balconies, were settling down enough for voices to be clearly heard, Velucci brought in the Cambridge public high-school choir to set a properly lofty tone for the fateful deliberations that were to follow under the scorching lights of the TV cameras. With appropriate fervor, the choir sang, "This land is your land."

No one has more succinctly described the essence of the evening than William Bennett, science editor of Harvard University Press, and Joel Gurin, a free-lance writer from California, did in a subsequent report they published in the *Atlantic Monthly.* "The most pained people in the room," they wrote, were the "scientists who had been summoned" by Mayor Velucci. "Physical discomfort and sweat were the least of their problems. Suddenly their careers hinged on the ability to defend highly technical biochemical work in plain English, a language some of them had not spoken for years."

When what Velucci referred to as the "Harvard team" of witnesses testified in their accustomed jargon about *E. coli,* P3, and EK2, the mayor told them to "please refrain from using your alphabet. We are laymen, and don't understand your alphabet. So spell it out for us."

The NIH had sent its most attractive emissary, Dr. Maxine Singer, from Washington, to defend the new recombinant DNA research guidelines. With his tongue very plainly in his cheek, Velucci brazenly asked her for her telephone number. When she gracefully finessed that one, he asked how she, a woman, had become a member of the "Harvard team." She politely stuck to her mission and explained why she thought the guidelines were sufficiently protective, but the mayor continued to repeat a question she couldn't answer: "Why didn't someone advise the mayor of Cambridge and the Cambridge City Council what was going on at Harvard?"

A Harvard official, in prepared testimony, made the mistake of attempting an answer in her stead. He said the Cambridge City Health Commissioner had been invited to some of the Harvard biohazards meetings. Virtually everyone else in the hall knew that Cambridge hadn't had a health commissioner for nineteen months!

During the long, often angry and openly hostile exchanges, which continued until almost 1:00 A.M., Velucci presented a resolution committing the council to ban all recombinant DNA research in the city, whether publicly or privately financed, for two years.

"If you pass that resolution," Professor Ptashne responded, tears dripping from his voice, "virtually every experiment done by members of the Biochemistry Department would have to stop, including experiments that no one, sir, *no one,* has ever claimed have the slightest danger inherent in them, namely recombinant experiments done under P1 conditions."

This was a serious misstatement of fact. A number of scientists had held that *all* recombinant DNA experiments were dangerous and should be prohibited. But Mayor Velucci wasn't sufficiently versed in the technicalities to recognize the mistake. Besides, he had a much more human response. It wasn't the least bit logical. But it was based on an admission he had extracted earlier from a Harvard witness, and anybody could understand it.

"When I was a little boy," he said, "I used to fish in the Charles River. And I woke up one morning and found millions of dead fish. And you, tonight, tell me that you dump chemicals into the sewer system that overflows into the Charles!"

The proposed two-year ban on all recombinant DNA work in the city remained on the council table when the crowd left. It would be voted on at a second council meeting two weeks thence.

Across the Atlantic in London, *The New Scientist* commented that Velucci had "certainly succeeded in scaring the wits out the Cambridge biologists." The quoted statement would have been more accurate if it had ended with the words, "the scientists of America." In laboratories across the country the Cambridge council meeting came to be described variously as "a circus," "an Inquisition," "another Scopes trial," and "another Lysenko affair." The last three of those epithets were obvious overreactions, for the scientists had escaped with nothing more serious than a vigorous verbal shaking.

One observer who understands the scientific mind well has since offered the following rationalization.

"The Harvard and NIH people thought that, because Washington said it was okay to go ahead, that was that. They were flabbergasted to discover that Al Velucci could have a noose around their neck in just a few days' time. Here's a guy raving and ranting about monsters and germs in the sewers and they had to stop what they want to do because of him. They just didn't understand."

The worst aspect of the scientists' failure to understand arose from the inability of many of them to recognize that they too were raving and ranting, with far less justification than Velucci had for his histrionics. They were seeking accountability only to themselves. He possessed formal responsibility for others, and was trying to stretch his limited authority to fulfill it. He was mayor—not the political boss of the city. The daily affairs of the place were run by City Manager James L. Sullivan. The policies under which Sullivan operated were determined by the City Council, a body consisting of nine members, each with one vote. Velucci's vote carried no more weight than the vote of any of the other eight members, none of whom was his political captive. His dramatics on the night of June 13 had the affirmative quality of establishing an extreme position from which it would be possible for the council to retreat and still have room to protect the interests of the populace.

Precisely that sort of strategic withdrawal occurred on July 7, when the council met again. Velucci's proposal for a two-year ban on all recombinant DNA experiments was shelved. In its place the council adopted, by a 5 to 4 vote, a compromise worked out by the chairman of the resolutions committee, David Clem, a hardheaded young Texan with some experience in business and a Ph.D. candidate at MIT. As a student, he was sensitive to the issue of academic freedom; as a public servant, he knew that no personal freedom can be absolute in a democracy. What Clem proposed was a three-month "good faith" moratorium—modeled after the Berg committee letter of 1974—on only P3 research. That allowed the thirty-odd Harvard scientists then engaged in recombinant DNA experiments at the P1 and P2 levels to continue their work uninterrupted, and did not interfere with MIT's application for a P3 facility. During the three-month interval, a Cambridge Experimentation Review Board would be appointed by City Manager Sullivan to study the situation and recommend a future course for the City Council to follow in regard to the recombinant DNA problem.

The action of the City of Cambridge has been widely misread as an isolated local happening. In actuality, it completed a full turn in a cycle of national politics. The cycle had begun in 1971, the year of Robert Pollack's transcontinental telephone call to Paul Berg.

For several decades before then, the cutting edge of scientific knowledge in the Congress abided in the Science and Technology Committee (originally the Science and Astronautics Committee) of the House of Representatives. During an International Science Policy Proceeding of that committee, in 1971, Nobel laureate James Watson drew attention to the speed at which research in molecular cloning was moving and warned:

> This is a matter far too important to be left solely in the hands of the scientific and medical communities. The belief that . . . science always moves forward represents a form of laissez-faire nonsense dismally reminiscent of the credo that American business if left to itself will solve everybody's problems. Just as the success of a corporate body in making money need not set the human condition ahead, neither does every scientific advance automatically make our lives more "meaningful." . . . A serious effort to ask the world in which direction it wishes to move [might find molecular cloning confronting] a blanket declaration of worldwide illegality. . . .
>
> [If] we do not think about the matter now, the possibility of having a free choice will one day suddenly be gone.

But when the work of Stanley Cohen and Herbert Boyer and their associates in 1972 began to make cloning look like child's play, the subcommittee of the House Science and Technology Committee, which should have been concerned with the new phenomenon, was headed by a congressman who was being bedeviled in his home district by critics of fetal research and its moral and ethical implications. Fetal research came under the general rubric of genetic engineering, as did gene splicing. So the politic choice for the subcommittee chairman seemed to be avoidance of public hearings on gene splicing. Called for instead was a studious review of the history of genetic engineering by the Congressional Research Service of the Library of Congress. Dr. James M. McCullough wrote the prescribed document, which was solidly informative and appropriately unprovocative but could not save the subcommittee chairman from defeat at the polls in the next election.

In the Senate, Edward Kennedy of Massachusetts had been engaged for some time in a prolonged inquiry into the listless way in which agencies of the Health, Education and Welfare Department, including NIH, monitored the self-regulatory devices that pass for controls over the manifold lines of research that NIH fosters. He had been disappointed by his findings and was looking about for an opening through which to inject a new spirit. Less than two months after the close of the International Conference on DNA Molecules, his Labor and Public Welfare Subcommittee on Health was holding public hearings on the promise and problems of gene splicing. Very early he was asking the head of NIH what reason there was to believe that voluntary controls over recombinant DNA would be any better enforced than controls on other areas of research funded by NIH in the past. Kennedy was also championing the right of laboratory workers to be protected from the consequences of experiments they had no say in, and the right of local communities to involve themselves in the regulatory process. By way of example, he had given sympathetic national exposure to such people as Stanford Professor Halsted Holman, who cited experience with California statutes to support his conviction that ordinary people can understand and respond responsibly to new developments in science when there is early and open opportunity.

Except for several forays by Kennedy, sometimes joined by New York Senator Jacob Javits and Senator Richard S. Schweiker of Pennsylvania, the larger social context in which recombinant DNA research should have been debated from the beginning had remained neglected except by a few broad-gauge scientists such as Chargaff and Sinsheimer. As far back as July 1974, the Berg committee had called upon the director of NIH to initiate experiments that would rigorously test, rather than merely guess at, the real risks of hybrid DNA molecule manipulation. HEW itself, in chartering the Recombinant DNA Molecule Program Advisory Committee in October 1974, had charged the committee first of all to formulate a program for such laboratory testing. Science for the People had urged a broad multidisciplinary approach to a testing program even before the International Conference on Recombinant DNA Molecules. The Statement of the Proceedings of that conference had argued a similar case in February 1975. And in May 1975 the Recombinant DNA Molecule Program Advisory Committee had appointed a subcommittee to design and oversee the experiments. But the subcommittee was allowed to atrophy and die of inactivity, and nothing of consequence happened until August 1976, the month immediately following the Cambridge City Council moratorium action.

The August 31 assemblage of NIH consultants has received almost no attention in popular accounts of the recombinant DNA controversy. But its pivotal role in the collective mind of NIH is evident in the care with which staff scientist Malcolm Martin took notes of the proceeding. There are ninety-three single-spaced typewritten pages. Every change of speaker is meticulously recorded. None of the speakers is identified—not even the one who opened the meeting by characterizing the assemblage as "basically an ad hoc group of consultants" whose assignment was "to try to give us a crash course in all of the things you need to know about infectious disease and were afraid to ask." Neither the need for nor the desirability of such total anonymity was explained.

The City of Cambridge was very much in the thoughts of the infectious disease specialists and enteric biology experts who filled the room. Martin's notes give this as the first exchange:

—"You are all aware of the molecular politics of recombinant DNA. Those from Boston are acutely aware . . . of how the reaction has flared up to the heights of glory, fanned by a number of Nobel Prize winners."

—"Just one."

The "one" became a subject of subsequent comment by a consultant who despaired of ever learning how to "cope" with him, and by another consultant who said he had spent an "immense amount of time trying to refute Wald point by point" and had found the task "hopeless."

An appalling cynicism about the intelligence of people who are not scientists was displayed again and again as the day wore on. One observation: "The public can't think in terms of multisteps and pathogenic mechanisms. The public has to think in terms of slogans and vague generalities and abstractions." Another observation: "As far as the heavy criticism is concerned, one doesn't do a scientific experiment, one does a political experiment. One does the experiment and finds nothing and makes a big deal about it, even though it may not make any scientific sense."

There was fearful talk of "housewives looking over your shoulder" in the laboratory. There was a desperate cry of resignation that "politically, the recombinant DNA business is lost in this country. I think it is hopeless; I absolutely believe that. I am going to fight to the last man, but I think it is hopeless. The real problem is [that] not all science should go down with recombinant DNA. At a certain point, I think we have to abandon ship. I am serious. Otherwise, what's going to happen is that everything is going to go down. All of genetics, which I care more about than recombinant DNA."

Just before the lunch break a calmer voice broke in: "If you talked to historians, they would say, 'Well, science has had too much money anyway. Why doesn't somebody allow me to study eighteenth-century wars?' There is a sense of entitlement in scientists, which seems to be very strong in molecular geneticists, that they should be allowed to do a large number of experiments the way they want to do them." But the dominant note of the day was struck once again, at another time, in what amounted to a threat: "Grass-roots' outrage at some point carries a lot of weight. . . . A few people can have tremendous impact and I think somehow it can be used in the other direction (viz., that scientists at some point are going to say, 'This is enough')."

None of this had very much relevance to the stated purpose of the meeting. According to the NIH moderator of the discussion (who, in Martin's notes, was as anonymous as anyone else present), the basis for the recombinant DNA research guidelines was 90 percent emotional and only 10 percent scientific. However, he said, "there is an immense amount of information that could put this whole thing into a much more satisfactory and widely acceptable frame of reference." This data, he went on, was in the hands of infectious disease specialists who had not previously been asked for their advice as to which forms of hybrid DNA manipulation were safe and which were not. What the NIH moderator neglected to say was that whatever failure there had been to consult epidemiologists was not due to the lay public. The responsibility rested first with the Berg committee and second with the Recombinant DNA Molecule Program Advisory Committee and third with NIH.

Martin's notes offered no summation of the thought content of the meeting, perhaps because the consensus was too diffuse. As long as the discussants held to the K12 strain of *E. coli,* there seemed little doubt among them that the chances of an epidemic rising from any hybrids were extremely low—except among occupants of hospitals. Deliberate attempts to turn the normally innocuous K12 into an active agent of disease for the purpose of designing a vaccine to fight dysentery had failed. The possibility of infection of those who handle K12 in the laboratory was conceded. That, however, for some unstated reason, was not considered a serious matter.

Deciding what experiments would conclusively demonstrate the unlikelihood of secondary contagion beyond the laboratory space was not an easy exercise, it turned out. Human subjects would have to be involved, it was agreed; how could their cooperation be obtained? The use of "green enzymes" was suggested—"not enough to make anybody rich." A thousand volunteers might be required; could that many be motivated by the green stuff? No one was willing to say so, although several did say that the results of such an experiment would be worth a million dollars to NIH. Even more difficulty was experienced with possible designs for a means of monitoring infection below the epidemic level: infections that would strike here and there for no apparent reason. That was the region where the questions of Ruth Hubbard and George Wald had proved most embarrassing, especially when the Legionnaire's Disease was in the background.

When the beleaguered consultants turned away from *E. coli* and its plasmids and phage to talk about the SV40 and polyoma viruses and their experimental use in prokaryote-eukaryote gene splicing, the prospects for public acceptance seemed even darker. "I don't think that the public concern is at all focused on those yet," one of the consultants ventured. "But [it will be] once you start saying, 'Well, I am going to create a new tumor virus, one that produces dozens of different kinds of tumors.' All you have to do is let that get onto the front page of the *New York Times* or the *Boston Globe* or something and you're in trouble."

Copies of the notes of that August 31 meeting were distributed to members of the Recombinant DNA Molecule Program Advisory Committee at their meeting in September 1976, and the committee called for a follow-up workshop on recombinant DNA risk assessment in the following spring. Until then, the primary focus of the increasingly rebellious scientists' resentment would be on the Cambridge-Washington axis, happenings at one end of which were obviously affecting happenings at the other end.

Those who originally thought that the Cambridge affair could be dismissed as an ordinary political ploy had already discovered otherwise. Having witnessed the bitter outbreaks of antagonism among the scientists, and knowing no reason to think that the scientists could agree on anything meaningful within the span of time he had to work with, City Manager Sullivan chose to name an Experimental Review Board consisting entirely of nonscientists. The task was highly explosive politically. He had eight places to fill in addition to the chairmanship, for which the new City Commissioner of Health and Hospitals, Dr. Francis Communale, had already been designated. Every geographical segment of the city had to be represented by those nine; Sullivan managed that aspect of the job first. Ethnic considerations came next; he fulfilled them by including an Irishman, a Yankee, an Italian, a Frenchman, a Jew, and a black among his appointees. The two sexes had to be treated equally: Sullivan gave four seats to women and four seats to men. Cambridge had two broad election-day influences to take into account—the Cambridge Civic Association and those who run for office independently of the CCA candidates; Sullivan chose a wealthy female from the CCA who had served on the City Council and a male independent who once had been mayor. To assure the Experimental Review Board of an unprejudiced source of knowledge of physical containment, Sullivan chose a construction engineer, one of the foremost high-rise building authorities in the country. To provide religious perspective, Sullivan named a nun who was not only a hospital nurse but an administrative functionary in the hospital. To represent business, a heating oil company manager was named. To speak for the poor, there was a black social worker. Sullivan completed the Experimental Review Board roster with a student of urban policy who had a degree in the philosophy of science, a physician who specialized in treatment of infectious disease, and a cousin of Velucci's—a civic activist who had a routine 9-to-5 job with the Carter's Ink Company.

It was August 6 before Sullivan could write the formal charge for the Experimental Review Board, and the first board meeting did not take place until August 26. Then it was discovered that only three of the nine members had ever met before. At the second meeting, on September 14, the board requested and the City Council granted an extension of the moratorium for three additional months. Harvard and MIT officials agreed to abide in good faith that much longer. At the end of that time, the board would report its findings to Health and Hospitals Commissioner Communale. To avoid a conflict of interest, Communale relinquished the board chairmanship to the vice-chairman, former Mayor Daniel Hayes, the heating oil company manager.

Twice a week after that, on Tuesday and Thursday, the Experimental Review Board met in the Cambridge Municipal Hospital boardroom until the report was turned over to Communale. Each meeting lasted at least two hours. The Tuesday meetings were executive sessions. The Thursday meetings were open to the press and public. Proponents and opponents of recombinant DNA research were heard on alternate weeks, except on the day when a five-hour-long mock trial was staged. Altogether, seventy-five hours were spent in testimony taking, followed by twenty-five hours of study and deliberation, illuminated by tours of the Harvard and MIT laboratories and a mock experiment that demonstrated all the steps in the recombinant DNA process.

Although advance provision had been made for minority reports, the findings that the board presented to Communale on December 21 were unanimous. On the board's behalf, City Manager Sullivan called a press conference for 7:00 P.M. on January 5. Velucci tried to recapture attention by summoning a council meeting for 6:30 P.M. Although the mayor presided, the council voted to recess during the press conference and the Experimental Review Board's recommendations stole the show.

The recommendations were three in number. One proposed that recombinant DNA research be permitted in Cambridge with adequate safeguards. But the NIH guidelines alone were not judged adequate. Additional protection was needed. Institutions in which the research was conducted should be required to prepare comprehensive procedure manuals for all levels of containment of biohazards. All laboratory personnel should be trained to minimize any accidents that might occur. The institutional biohazard committees mandated by the NIH guidelines should include members from a variety of disciplines, representation from the biotechnical staff, and at least one community representative with no institutional affiliation. P3 experiments should be conducted with NIH-certified host-vector systems of at least the EK2 level (a vector carries foreign DNA from host to host). Host organisms used in experiments should be screened for purity. Organisms produced by experiments should be tested for resistance to antibiotics commonly used in therapy. Survival of experimental organisms in the bodies of laboratory workers and the possible escape of those organisms into the environment should be monitored. And Cambridge should appoint its own biohazard committee to oversee all recombinant DNA research in the city, inspecting research sites when and where necessary.

Recommendation number two called on the City Council to enact an ordinance making any violation of the guidelines and proposed additional regulations a health hazard to the city.

Recommendation three urged the council to appeal to Congress to adopt federal legislation applying the NIH guidelines to all recombinant DNA research whether publicly or privately funded, establishing a registry of all recombinant DNA research laboratory workers, and providing funds for research into means of monitoring the fate of experimental organisms.

The council showed its enthusiasm by enacting the requested ordinance, fixing fines of $200 a day for each day of violation, and providing authority to shut down offending laboratories. The appointment of a Cambridge Biohazard Committee was also authorized. But when City Manager Sullivan wrote to NIH requesting financial help to pay the city's share in monitoring the safety of experiments, he got a noncommittal reply.

Velucci thought that he and the other council members deserved a Nobel Prize for what they had done. Or, if not that accolade, then some other international honor. He was, therefore, chagrined not to receive an invitation to the National Academy of Sciences in Washington when the academy finally scheduled the recombinant DNA forum that it should have held in 1973. After all, former Cambridge Mayor Hayes, the Experimental Review Board chairman, had been invited. Whether he really had other business in Washington on March 7, 1977, the forum's opening day, or not,

Velucci said he did; and he showed up unexpectedly at a pre-forum press conference arranged by the academy's public information officer, Howard Lewis. In a querulous voice, the mayor demanded to know why he had been excluded from the guest list.

Ordinarily, his tactic might have had a good chance of making a headline. But the academy's quota of headlines that day had been preempted by a man named Jeremy Rifkin, who headed a volunteer citizen's group called the People's Business Commission. Before Velucci reached the scene, Rifkin had sent the academy and the Washington news media identical copies of a telegram threatening to break up the forum before it could get rightly started unless he was given time at the opening session to state the People's Business Commission's belief that the forum was a fraud.

The forum really was a fraud in one important respect. The academy was pretending to hear and respond to all points of view equally, a pose that could not survive skeptical examination. For the academy's Life Sciences Assembly had already voted to support the NIH guidelines as they stood. Whatever had to be said about this had better be said openly than leaked underground. So Rifkin was allowed to stage his curtain raiser. He arranged his followers in semicircular phalanxes in the academy's stately Great Hall, where they clapped their hands in unison and chanted, "We shall not be cloned!" Velucci was completely forgotten as Rifkin extorted an agreement that every scientist who spoke would identify the sources of his income. With all his effrontery, the mayor of Cambridge had never put the Harvard and MIT researchers in such an unwelcome spot.

A certain poetic injustice permeated events that followed. Because of his irrepressible flamboyance, Velucci, the man who had the courage to prod the Cambridge City Council (made up of a housewife, a student, a security guard, a welfare mother, a pharmacist, a college administrator, a rubbish disposal contractor, a court clerk, and a tax collector) into taking action that the Congress of the United States had avoided, was passed over when the recognition he dreamed of came in sight. Later in that month of March, Experimentation Review Board Chairman Hayes and Council member Clem were the ones who spoke for the city at hearings conducted by Congressman Paul G. Rogers's Subcommittee on Health and the Environment of the Committee on Interstate and Foreign Commerce of the House of Representatives. And it was Hayes and Clem again who testified before the Kennedy Subcommittee on Health and Scientific Research of the Committee on Human Resources of the Senate. During the House hearing, Congressman Dale Bumpers of

Arkansas, author of a bill that would put strict restraints on recombinant DNA experiments, emphatically thanked the witnesses for what they and their colleagues in Cambridge had done. The Governor of Massachusetts, Michael Dukakis, appeared at the Senate hearing and praised the Cambridge lawmakers for setting a national example.

The purpose of the Kennedy hearing was to solicit variant views of three pieces of legislation that had been introduced in the Senate earlier. One of the three was the Bumpers bill, which the Arkansas senator offered in February. It provided for licensing of all recombinant DNA research by the Secretary of Health, Education and Welfare, inspection of sites of experiments, levying of jail terms up to one year, and fines up to $10,000 for each day of violation of the license. The bill also held experimenters strictly responsible for any damage their work might cause. A second bill, introduced in March by Senator Howard Metzenbaum of Ohio, provided equally tough penalties for license violators but assessed no liability for damage; it called also for establishment of a thirteen-member national commission to study the need for further legislation. Kennedy himself had introduced the third bill on April 1, on behalf of the then new administration of President Jimmy Carter. That bill gave virtually unlimited authority to the HEW Secretary, to the extent of exempting from regulation any experiment which in his opinion did not constitute a hazard or was already covered by existing law. Kennedy didn't like the loopholes in that bill, and talked of taking the authority over recombinant DNA research away from HEW and putting it in the hands of a national commission whose majority would be made up of laymen. He expressed the hope of getting the legislation moving by May 15. But he grossly miscalculated the lengths to which the scientists would go to maintain their freedom to experiment as they saw fit without regard to the wishes of the people.

26

THE FIRST indication of the presence of a subtle form of intellectual terrorism came from the University of California Medical School at San Francisco, home of the famous Boyer laboratory. On May 31, 1977, Boyer wrote a letter to Ralph Backlund, executive editor of the *Smithsonian* magazine, complaining that several statements in an article published in the June issue of *Smithsonian* (copies of that magazine are delivered to homes of subscribers during the two weeks preceding the first day of the month of issue) had been inaccurate. Boyer's tone suggested more hurt and disappointment than anything else. But on June 2, Dr. David W. Martin, Jr., a physician who then was chairman of the Biosafety Committee of the University of California, San Francisco, also addressed a letter to Backlund concerning the same article that offended Boyer. Martin's tone was shrill and intimidating. "It is unfortunate for all," he said, that the writer of the piece "was not . . . a thorough investigative reporter. It is clearly an inaccurate, unnecessary, damaging article and reeks of sensationalistic journalism"; the author's reporting of "unfounded . . . rumor," and the *Smithsonian*'s "negligence in allowing that article to be published unverified have undoubtedly generated much unwarranted anxiety and mistrust in an arena of interaction between lay society and scientists already jeopardized by a climate of suspicion."

The author of the questioned article, Janet L. Hopson, was an established science writer with a sound reputation among the scientists whose work she reported. She had earned a position on the Washington staff of *Science News.* In that capacity she had covered the Berg committee press conference at the National Academy of Sciences in July 1974 and then the international conference at Asilomar in February 1975. From what she heard there she had become convinced that recombinant DNA was the "hottest science story of the seventies." Having reached that conclusion, she determined to make herself expert in the subject by studying a recombinant DNA lab from the inside. Her qualifications for the task impressed both the Council for Advancement of Science Writing and the National Institutes of Health. They together gave her a special research grant. She used the grant as a free-lance. Only at the end of her self-assigned job did she sell a report of her observations. The publisher of the popular science magazine she sold to was the Smithsonian Institution, an American science fixture whose roots go much deeper than those of the National Academy of Sciences. As Martin said at the close of his letter to Backlund, "The Smithsonian Institution and therefore its publication have been esteemed."

Whatever trangressions Hopson had committed to deserve Martin's attack on her credibility and her publisher's integrity must have been serious indeed, or so one would suppose.

What did Hopson write in the *Smithsonian?* It was a layman's account of ninety-five days spent as a worker in Boyer's lab. The scene was Boyer's lab by accident. Hopson had gone first to Professor Falkow at the University of Washington in Seattle, whom she knew well. Falkow felt that her study would be more meaningful if it took place closer to the center of recombinant DNA activity, and he sent her to Professor Cohen at Stanford. Cohen has been an outspoken opponent of too much popular involvement in scientific research, especially when the involvement results in government regulations of laboratory research. He sent Hopson to Boyer. Boyer accepted her (and her insistence that he could not read her manuscript prior to its publication) because, he told Backlund, "I could see that this might be one way to promote a better understanding between science and society."

"What sort of person, in what kind of Strangelovian laboratory, spends his time making genetic chimeras? What do these half-breeds [the chimeras] look like? What, for that matter, does DNA look like?" Those were some of the questions Hopson wanted to answer. And the early paragraphs of her report in the *Smithsonian* were filled with expressions of pleased surprise that Boyer, one of the "eleven luminaries" who signed the "Berg letter," looked like a curly haired boy though he had just turned forty. He jogged two hours a day, climbed mountains to hold his weight down, and used such exclamations as "Wow!" and "far out!" She was captivated by the Olympian quality of the view from his windows, looking down as they did from Mount Sutro in two directions across the Golden Gate Bridge to the sea. The most fearsome approach to a chimeric animal she saw on the premises was a small black mouse sitting in a wire cage. She was struck by the informality of Boyer's research staff, and by its international dimensions. She was thrilled to be allowed to take a small part in an experiment, one in which restriction enzymes were removed from their natural host, a plasmid that was not very prolific, and transferred to another plasmid that *E. coli* would make many copies of.

After she had been part of the lab's daily life for a month and a half, she felt free to mention, "over hamburgers in the cafeteria," that she had just read "yet another newspaper story on the growing public concern about recombinant DNA research." She ticked off some specific events: the Cambridge moratorium, the debates in the New York and California legislatures, and the bills in Congress.

"Most of those at the lunch table" felt that no laws were necessary, Hopson wrote. "The work is totally harmless," she quoted one as saying, "and controls are holding up vital research. We have enough growth hormone genes upstairs right now to clone for a year, but with the guidelines we can't do a thing." "It's ridiculous," another of her table mates added. "Other researchers right here are doing dangerous studies—like infecting human tissue with monkey viruses—without any special containment. But we can't even use human tissue in our work without expensive containment equipment and special organisms that aren't even available yet." Hopson quoted a third person as joining in: "The bacteria we use [the K12 strain of *E. coli*] are perennially sick and live only briefly outside the test tube." This person, Hopson wrote, "urged the first person to go ahead with his growth hormone studies anyway. 'We're not from the U.S.,' he said. 'No one has to know if you go ahead a little early. You can repeat the experiments later and publish them from your own country.' This suggestion was considered and politely declined."

After that lunch talk, Hopson wrote, she "brushed up on the NIH guidelines and watched day-to-day lab procedures more closely." She was involved in P1 work, which, she explained, "requires no special equipment or organisms but does demand good microbiological technique." She now observed that the researchers around her "rarely wear white laboratory coats (suggested by the guidelines) and they smoke, eat, and drink in the lab while experiments are in progress (expressly discouraged) [by the guidelines]. I also watched people sucking recombinant organisms into glass pipettes with their mouths instead of [with] the recommended rubber bulbs, and saw work surfaces decontaminated only sporadically—instead of daily, as required. Half of the researchers here follow the guidelines fastidiously; others seem to care little."

A Ph.D. candidate from another university came into Boyer's lab and "did experiments with a disease-causing bacterial strain," Hopson noted. These experiments, she said, "would seem to require a higher level of containment" than the P1 afforded, "probably a costly type of ventilation hood." Someone asked the experimenter what level of containment was specified by the guidelines. "Damned if I know," Hopson quoted him as replying. Someone else asked him what level of containment he used in his own lab. Hopson described his reaction: "He shrugged 'I never looked it up.'" Hopson thought this a cavalier attitude, and asked her teacher and coach, Ray Rodriguez, what he thought. "Ray . . . reminded me," she wrote, "that there have been no accidents with recombinant organisms to date, either here or elsewhere, and if anyone is at risk it is themselves."

Almost three weeks later, Hopson reported, she discovered that "the high containment (P3) lab upstairs was shut down by the university's biohazard committee for a few days this week. I was dismayed to hear people joking about the closure and the messy conditions that precipitated it. Among the young graduate students and postdoctorates it seemed almost chic not to know the NIH rules."

On the ninetieth day of her ninety-five day assignment, Hopson was in Boyer's office when he got a telephone call from a science reporter in Washington who thought he had inside information on the first biohazard accident in a recombinant DNA lab. "He was eager to break the story," Hopson recalled, "and had called Boyer in hope of confirmation." The story, Hopson said, was that a "bacterial virus that couldn't be killed—a true Andromeda strain—had been created by accident at the San Francisco laboratory."

Hopson wrote that she heard Boyer insist, "No, that's not true. It didn't happen that way at all. You're not listening. . . . Look, I'm very busy and you're just wasting my time by arguing." The real story, Boyer assured the reporter, "was not so exciting." Hopson continued: "A sloppy technician had spilled some growth medium contaminated with ordinary bacterial virus and had failed to clean it up properly."

After hanging up the phone, Boyer brought out a stack of letters from people opposed to recombinant DNA research. Hopson said he called it "hate mail." Her comment: "The idea that observers from the community or their legislative representatives could assert control over daily laboratory routine seems foreign to the sober, scholarly people who study bacterial genes. A revolution in the relationship between science and society is already well under way, yet attitudes and safety procedures, at least for some workers, still lag behind their new public accountability."

What in Hopson's report did Boyer object to? "Two factual misrepresentations . . . cannot be ignored," he told Backlund. First, "it is not true that the visiting scientist was working with a disease-causing strain. . . . It had been experimentally demonstrated that in this case it could not cause any pathological response." Second, the P3 laboratory "had not been shut down by the campus Biohazard Committee"; the committee had simply delayed its approval of the lab for P3 experiments. Finally, while it was "true that in theory the NIH guidelines are extensively criticized (by both sides of the controversy), in practice they are followed seriously. Unfortunately, the article gave the erroneous impression that this was not the case."

And Martin's attack on Hopson and the *Smithsonian*—what basis was there for it? He said he had a "soft criticism" of Hopson's description of laboratory habits and practices. "Although true," he said, her remarks were "unfortunately placed in a demeaning context." Martin also objected to Hopson's commenting on the "Andromeda strain" rumor "without having stated that it was unfounded." Further, there was a "very serious" misrepresentation of fact about the P3 lab. "Ms. Hopson states that . . . the high containment (P3) lab upstairs was shut down by the university's Biohazards Committee for a few days this week," Martin wrote. As chairman of that committee, he said he could "unequivocally state that such did not occur." He had inspected the lab before its approval for P3 work, and approval had been refused until packing materials from recently unpacked supplies were removed from the facility and a new lock with only three available keys had been put on the door. Before the lab was opened for P3 work in 1976, he insisted that the "facility had never been used for an experiment."

"Clearly Ms. Hopson has her 'facts' confused," he resumed. "I have no patience with . . . her not verifying her information. A charged issue such as DNA recombinant technology deserves more thorough investigating . . . than she must be capable of."

Later events would vindicate Hopson. They would show that the Biosafety Committee chairman who accused her of failing to do her homework knew at the time that experiments conducted in the P3 lab had in fact violated NIH guidelines. In January 1977 a post-graduate associate from abroad, working with the knowledge and approval of the head of UCSF's Biochemistry and Biophysics Department, had made ten clones of rat insulin genes with an *E. coli* plasmid while the safety of the plasmid was still being tested by NIH. Biosafety Committee Chairman Martin had learned of the violation in May and had discussed it then with the Dean of the UCSF Medical School. On June 3, the day after he signed the letter excoriating Hopson and the editors of the *Smithsonian* for "inaccurate . . . sensationalistic journalism," Martin shared his prior knowledge of the violation with other members of his committee. How the committee was used to cover up the facts after it got the news, and how at least one NIH official took part in the cover-up, will be told later in these pages. Note should be taken here, however, of a striking coincidence in time between the smearing of Hopson and the beginning of an intense campaign within the scientific community to persuade the Congress and the American people that scientists can effectively govern themselves without legislative constraints. Was the coincidence entirely accidental? Or did some strongly opinionated person or persons act in anticipation of probability that the gene splicers' propaganda could not flood the communication channels if Hopson's account of contempt for discipline in the UCSF laboratory should first gain unchallenged acceptance in official and popular minds?

27

JUNE 3, 1977 was the day when members of the American Association for the Advancement of Science received their copies of the June 10 issue of *Science*. Included was a letter from Professor Harlyn O. Halvorson, head of the Rosenstiel Basic Medical Science Research Center at Brandeis University and president of the American Society for Microbiology (ASM).

"There is at present no demonstrated evidence that microorganisms containing recombinant DNA molecules are hazardous," the letter said. "However, because federal legislation concerning the production and use of recombinant DNA molecules is under active consideration, and in the event that such legislation is passed," Halvorson wished to acquaint the AAAS membership with a unanimous action by the ASM council taken the month before.

If accepted as a model for emulation, that action would deny the local option implicit in the Cambridge City Council ordinance and pull all the teeth that Senator Kennedy's staff people were known to be putting into legislation intended to regulate recombinant DNA experiments. The ASM Council held that all authority over recombinant DNA research should rest in the HEW Secretary, who should be assisted by an advisory committee and should delegate as much responsibility as possible to local institutional biohazard committees. Institutions should be licensed, but not individual scientists, nor should individual scientists be bonded or otherwise held liable for damages their experiments might cause. Furthermore, the ASM Council would exempt all P1 experiments—that is, at least one-third and perhaps as much as two-thirds of all ongoing recombinant molecule research—from even the NIH guidelines.

The significance of Halvorson's letter was not explained in the news columns of *Science,* although it deserved to be. For the letter represented ASM's first involvement in the recombinant DNA controversy. It will be recalled that the Berg committee at its organization meeting in 1974 had considered the desirability of but decided against liaison with ASM rather than with the National Academy of Sciences. Halvorson subsequently had tried and failed to obtain an invitation to the International Conference on Recombinant DNA Molecules in February 1975. ASM's exclusion from the deliberations continued until May 1976, when NIH Recombinant DNA Molecule Program Advisory Committee Chairman Stetten appeared at the annual ASM convention to ask the society's comment on the gene-splicing guidelines. An *ad hoc* ASM study panel was then appointed. At the end of September 1976 the panel reported its judgment that the guidelines provided "proper and adequate procedures for handling hazardous bacteria." In May 1977 the ASM Council passed resolutions supporting the panel's findings in the manner Halvorson described in the letter to *Science.*

The ASM had 25,000 members. With that weight behind them, the recombinant DNA research enthusiasts felt that they had regained their original initiative and should lose no time in finishing off the opposition. A scientific lobby, the like of which had not been seen since the bill establishing the National Academy of Sciences was sneaked through a lame-duck Congress in Abe Lincoln's day, was hastily organized.

The ASM had 25,000 members. With that weight behind them, the recombinant DNA research enthusiasts felt that they had regained their original initiative and should lose no time in finishing off the opposition. A scientific lobby, the like of which had not been seen since the bill establishing the National Academy of Sciences was sneaked through a lame-duck Congress in Abe Lincoln's day, was hastily organized.

Prominent among the fomenters was Harvard bacteriologist Bernard Davis, who had contended from the beginning that critics of gene-splicing were grossly exaggerating the seriousness of the potential hazards of recombinant DNA research. Davis took every opportunity to rub salt on the psychological bruises of Halvorson, who had been ignored by Congressman Paul Rogers' subcommittee in the House of Representatives and had managed to get onto the Kennedy witness list in the Senate only after threatening to flay the Massachusetts legislator at a press conference. (Halvorson's reaction to his circumstances was typical of the disoriented perspective of the lobbyists; instead of blaming the scientists who had snubbed him and the society he represented, he directed his resentment at a public office-holder.) The lobby wasn't as powerful as it looked or sounded, for the scientific community as a whole was as deeply divided as ever on the issue of public responsibility. When, for example, an emergency meeting of biological societies was called to drum up reinforcements, the vocal support of twenty organizations was proclaimed. But the biggest of all scientific societies, the American Association for the Advancement of Science, refused to stampede. To paper over that important defection, AAAS board chairman William D. McElroy, chancellor of the University of California, San Diego, was asked (and agreed) to join the lobby as an individual along with nine other "prominent individuals" who spoke only for themselves. What the freedom fighters lacked in numbers, however, they made up for in noise, academic degrees, and clever (if not always highly principled) stage managing.

For dramatic effect, the first objective was to generate an image of reversal of a mistaken idea at its source. This meant going back to the Gordon Research Conference on Nucleic Acids, from which in 1973 had come the earliest open expressions of concern over the possible dangers of hybrid DNA molecule experiments. The reversal had to be an illusion, for the original questioning had been a spontaneous response to Herbert Boyer's report on the new gene-splicing technology, and there was no spontaneity in the lobbying of 1977; everything about it was contrived.

Dr. Frederick Blattner, a University of Wisconsin researcher who was impatient with delays in approval of the "disarmed" phage he had designed for use with Curtiss's χ1776 bacterium, took copies of the Kennedy bill and other proposed legislation with him to the 1977 Gordon Conference. At the opening session on Monday morning, an announcement was made: An "informational" meeting on the political situation would be held later. Prominent among the "informants" was Dr. Alexander Rich, of MIT, a member of the governing board of the National Science Foundation, which professes not to involve itself in politics. Other instigators were David Botstein of Harvard and Howard Goodman of the University of California in San Francisco. Several drafts of a letter to Congress were worked on for several days, and the final draft was endorsed by 173 conferees—86 percent of those in attendance.* Harvard's Walter Gilbert signed an introductory statement that appeared with the letter when it was published in *Science.* The statement said the conferees agreed that "regulation beyond simple enforcement of the NIH guidelines is unnecessary, and many expressed the view that less regulation would suffice to guard against any hypothesized dangers."

Without offering any documentation to support the statement, the letter itself said, "This meeting made apparent the dramatic emergence of new fundamental knowledge as a result of application of recombinant DNA methods. On the other hand, the experience of the last four years has not given any indication of actual hazard. Under these circumstances, an unprecedented introduction of prior restraints on scientific inquiry seems unwarranted. . . .

"We are concerned . . . that legislative measures now under consideration by Congressional, state and local authorities will set up additional regulatory machinery so unwieldy and unpredictable as to inhibit severely the further development of this field of research. We feel that much of the stimulus for this legislative activity derives from exaggerations of the hypothetical hazards of recombinant DNA research that go far beyond any reasoned assessment."

The lobbyists benefited from a supporting road show in June. This was the risk assessment workshop recommended by the NIH Recombinant DNA Molecule Program Advisory Committee at its September 1976 meeting. The workshop was held in the bar of a hotel in Falmouth, Massachusetts. By law, anyone who wished to attend had a right to be there. In actuality, a policy of calculated discouragement of possibly skeptical observers was followed. My own experience illustrates the deviousness of the method. I telephoned the office of recombinant DNA activities at NIH to ask for details of the workshop. Dr. Gartland, the head of the office, was out of town. His assistant

* Dieter Soll, Maxine Singer's co-chairman of the 1973 nucleic acids conference, was among them.

told me the workshop sessions would be closed to the public. I felt reasonably sure that was not true. So I called Dr. Wallace Rowe, a prominent NIH staff scientist who sat on the Recombinant DNA Molecule Program Advisory Committee of NIH. I was right, he said, the workshop was open to the public. "But," he cautioned, "I don't know how the fire laws will be interpreted." Obviously, he was referring to the restrictions local fire departments place on the number of persons permitted in public rooms of stated size. "You are planning the workshop in a small hotel room?" I asked. "Yes," Rowe replied. "What do you think the capacity of the room is?" I asked. "About fifty people," he said. "And you have already invited forty-five people?" I persisted. "Yes," Rowe said. I learned later that the bar at the Falmouth hotel holds many more than fifty people.

Afterward, a member of the workshop steering committee told me the committee had agreed in advance that any summary statement or report would be written only after all committee members were consulted. No consultation ensued, but the workshop moderator, Tufts University Professor Sherwood L. Gorbach, afterward put "my own evaluation of the discussions" into a letter to NIH Director Donald Fredrickson. His justification for acting independently was that an "important consensus was arrived at . . . which I felt was of sufficient interest to be brought directly to your attention."

What since has come to be called the "Gorbach report" said that forty-five specialists "from the United States and abroad, with expertise in clinical infectious diseases, enteric bacteriology, epidemiology, gastroenterology, endocrinology, immunology, bacterial genetics and animal virology" had "arrived at unanimous agreement" that *E. coli* K12 was not normally an epidemic pathogen and could not be converted into one by "laboratory manipulations with DNA inserts." It was also the "consensus of the group" that the possibility of transferring plasmids from *E. coli* K12 to wild-type bacteria in natural surroundings was "extremely unlikely." "As a result of these discussions," Gorbach said, "it was believed that the proposed hazards concerning *E. coli* K12 as an epidemic pathogen have been overstated. Such concerns are not compatible with the extensive scientific evidence that has already been accumulated, all of which provides assurance that *E. coli* K12 is inherently enfeebled and not capable of pathogenic transformation by DNA insertions."

That the "Gorbach report" was hurried into print in *Nucleic Acid Recombinant Scientific Memoranda,* a newsletter published quarterly by NIH, was evident on the newsletter's face. Publication occurred in the issue that was supposed to close on June 15. The Gorbach letter was stamped "Rec'd June 14," although the workshop did not take place until June 20–21.

From the NIH newsletter the "Gorbach report" quickly got into the news colums of the *Washington Post* and onto the editorial page of *Science.* One more argument for weakening controls on hybrid DNA molecule experiments had been set afloat at a politically opportune moment in a medium that explicitly disclaimed standing as scientific literature. Observers who asked for a copy of the transcript of the workshop discussions were told it wasn't available. No one purported to know when it would be available. No date could be obtained for the promised formal publication of its substance in a refereed scientific journal. Most interesting of all, no announcement was made of the risk assessment experiments that were decided on, although Gorbach said they "should receive the highest priority."

Only after the case against recombinant DNA control legislation had spread through congressional cloakrooms did dissents from Gorbach's view of the workshop begin to be heard from other scientists who had attended. Richard Goldstein, an assistant professor at Harvard Medical School and a member of the workshop steering committee, wrote to NIH Director Fredrickson on August 30, saying that there had been "several serious misinterpretations by the press." For example, "though there was general consensus that the conversion of *E. coli* K12 itself to an epidemic strain is unlikely (though not impossible) on the basis of available data, there was *not* consensus that transfer to wild strains is unlikely. On the contrary, the evidence presented indicated that this is a serious concern."

Goldstein went on to say that the Falmouth workshop should be considered only as an "excellent beginning"; there should be other workshops and they should draw on a "much broader sector of the scientific, medical, and environmental committees. The meetings should be announced, allowing those who believe they have something to contribute to be heard. The strength of our scientific tradition lies in its open, democratic, nonexclusive nature."

Neither Goldstein's letter nor a concurring one—in which another Falmouth workshop participant, Associate Professor Bruce R. Levin, of the University of Massachusetts at Amherst, expressed opposition to "relaxation of the current NIH guidelines on recombinant DNA research or of the efforts to enforce them"—were given the benefit of the NIH propaganda machine and its volunteer adjuncts in the gene-splicing fraternity.*

Lobbying against the recombinant DNA research control legislation continued into July. Scientists attending another Gordon

* Months later, NIH implicitly admitted its distortion of the news by publishing the Goldstein and Levin letters along with the Gorbach report in a successor to *Nucleic Acid Recombinant Scientific Memoranda.*

Research Conference—this one concerned with biological control mechanisms—issued a statement obviously patterned after the letter written from the Gordon Research Conference on Nucleic Acids in June. The pending legislation on recombinant DNA research was "unwarranted and imprudent," the echoing communication said. "The bills presently before Congress would impose unjustifiable controls over many important areas of scientific inquiry. The premise on which these bills are [sic] based derives from exaggerations of hypothetical hazards that go far beyond reasoned assessment. The experience of the last four years is important. Despite a vigorous search to identify the degree and nature of any actual public health or environmental hazard, no indication of actual danger has been uncovered. Many conjectured dangers have been shown not to exist. It is evident that dramatic advances in fundamental knowledge have already emerged from application of these techniques. We are concerned that the legislative measures will set up a bureaucratic machinery so inflexible and unwieldly as to severely inhibit the development of many fields of knowledge."

The attack on the Kennedy legislation climaxed on August 1, when an accidental conjunction of circumstances made it appear that the American Association for the Advancement of Science, the largest scientific organization on earth, was pouring its intellectual resources into the lobbying effort. What really happened that day is not very well understood by very many, so it may be instructive to take a moment to elucidate. One of the circumstances had its roots in the appointment of William Carey as executive officer of the AAAS a few years earlier. Carey was superbly qualified for the job. Through the administrations of a number of presidents, both Democratic and Republican, he had been the top science expert in the White House budget office. But he was not a scientist, and his grasp of congressional intricacies did not entirely match his intimate knowledge of the executive departments and agencies of the government. Soon after accepting the AAAS appointment, he determined to correct these two shortcomings. It did not take long for him to see that he could do so in one move, by adding a young physicist, Dr. Thomas Ratchford, to his office. Ratchford had demonstrated unusual competence on the staff of the House of Representatives' Committee on Science and Technology, whose influence within the Congress has already been alluded to.

With Ratchford at his side, Carey was ready to put the immense reserve of AAAS knowledge at the disposal of the nation. Entirely apart from the recombinant DNA controversy, it was clear that Congress could use all the sound scientific advice it could get. Through Ratchford, Carey offered to perform an advisory function for any legislative committee that wished it. If the legislators would be ready with questions, special teams of AAAS members would appear to answer them as best they could on an unbiased, nonpartisan footing. And at the beginning of the last week in July 1977, word came from Capitol Hill that serious questions had arisen about recombinant DNA, questions that needed answers before another week began.

Carey had expected help in the new AAAS undertaking from AAAS board chairman McElroy. But McElroy was too busy in late July to accept the task, and he delegated it to a University of California professor who was writing a book on recombinant DNA. From his researches, he had reached the conclusion that the public was weary of the subject. For him, the steam was gone from the long and bitter debate. The lobbyists had triumphed. He said as much in conversations with news gatherers.

Whether or not the professor's perception of the situation affected the makeup of the AAAS team he recommended, the team in fact was lopsided with recombinant DNA lobbyists. When Carey saw the list of names on Sunday, July 31, he recognized that the AAAS could be in trouble. To head it off, he made a last minute appeal to Jonathan King, the Science for the People spokesman, in Cambridge. King agreed to join the team to provide some balance, but he had to be back home in Cambridge by Monday afternoon to keep a prior commitment. Carey strengthened the team further by adding two eminent figures—former Connecticut Congressman Emilio Daddario who had just resigned as director of the Congressional Office of Technology Assessment, and *Science* editor Philip Abelson—in the expectation that their prestige would hold the team within the limits of impartiality fixed by AAAS tradition.

Things did not turn out as Carey had hoped. The lobbyists on the team went in for the kill. Instead of waiting for questions, they immediately began haranguing their point of view. Within hours, Carey was getting negative feedback by telephone. So was at least one university, hundreds of miles away; it reacted swiftly, advising key personnel in Washington by wire that the recombinant DNA researcher from that school's faculty who had insulted the intelligence of the Congress should not be accepted as a spokesman for the school. At the end of the day, neither Ratchford—who had accompanied the AAAS team and observed its bizarre behavior—nor Carey could tell how badly AAAS credibility had been eroded by the episode. One point was clear: Jonathan King, theretofore assailed by many as an unprincipled radical, went home with high marks.

Misleading as the lobbyists' tactic had been, it worked. On the very next day, Wisconsin Senator Gaylord Nelson introduced an amend-the Kennedy bill. It gave the lobbyists virtually everything wanted, including exemption of P1 experiments from regulation even by the NIH guidelines. As testimony to the correctness of his position, Nelson put into the *Congressional Record* a verbatim reproduction of an article published in the July 31 issue of the *New York Times* under the byline of *Times'* science editor Walter Sullivan, who had written little about the recombinant DNA controversy up to that time. The Sullivan article began:

> One of the tragedies of modern science was the destruction of Soviet genetics during the 1930s and 1940s, in effect by government decree. It was decided by the leaders of the Communist Party that Trofim D. Lysenko was right and that classical Russian geneticists were reactionary and disloyal. Nikolai I. Vavilov, widely recognized as one of the world's leading plant geneticists, and many others were exiled. Vavilov apparently died in the labor camp at Magadan on the Sea of Okhotsk. Lysenko's ideas were applied to agriculture, resulting in near disaster.
>
> Now it is being argued that in the United States steps are about to be taken toward similar suppression of genetic research, though the motives differ. The concern is that legislation pending both in Albany and in Washington would establish elaborate—some researchers say stifling—controls over research involving manipulation of the key molecules in genetics: those of DNA (deoxyribonucleic acid). Such altered molecules are known as recombinant DNA.

In nineteen succeeding paragraphs, Sullivan recited the pro-recombinant DNA lobby's line with no critical observation of his own until the sixth from the last paragraph, which began: "The parallel with Lysenkoism . . . seems questionable."

As a stratagem of parliamentary procedure, Nelson's amendment was an effective strangulation device. It required debate of Nelson's ideas before Kennedy's ideas could reach a vote. The situation was further entangled by a redistribution of Senate committee jurisdictions that had been agreed on by the previous Congress. One of the shifts of responsibility gave the Senate Committee on Commerce, Science and Transportation legislative as well as oversight authority over science, engineering, and technology research and development policy. As chairman of the Subcommittee on Science, Technology and Space, Illinois Senator Adlai Stevenson IV responded to the anti-Kennedy lobby's pressure by promising to examine the "effect of proposed legislation to regulate recombinant DNA research on the conduct of basic research and the freedom of scientific inquiry." On September 21, Stevenson announced that he would hold hearings in

November on the "implications of such a precedent," which, he said, "must be handled with great care."

28 ON SEPTEMBER 27, 1977, Kennedy abruptly withdrew his support from the bill he had introduced on April 1. He announced the decision in a talk to the American Medical Writers Association in New York City. He said he would suggest a "compromise legislative approach" instead, one that would extend the NIH guidelines to cover "all parties conducting recombinant DNA research" for one year. He would also "help to form a National Recombinant DNA Commission" that would examine the situation for nine months and then decide "whether permanent legislation is needed." Scientific and public interest groups would be asked to nominate candidates for commission membership, and a final selection from those candidates would be made by an independent group headed by Dr. David Hamburg, president of the Institute of Medicine.

Kennedy left the medical writers convinced that he had not changed his mind about the inevitability of greater public participation in "critical policy decisions in science." On the contrary, he told them, the "recombinant DNA debate symbolizes" entrance into a "new era" in which the "blank check science policy of the 1950s and 1960s" will be abandoned in favor of application of the "principles of public accountability." "This new concept . . . has unsettled many members of the scientific community," Kennedy declared. "They see it as a threat to the freedom of scientific inquiry, or as the forerunner of government controls over science. But the truth is just the opposite. The greatest potential threat to science is the lack of public understanding and public participation."

Why, then, had Kennedy backed away from the DNA research control bill his Senate subcommittee had already approved? Because, he said, "the bitterness of the debate has obscured the constructive elements that are evolving." "The emotional atmosphere" had become too highly charged. Besides, the "scientific facts about recombinant DNA are in a state of flux. The information before us today differs significantly from the data available when our committee recommended the pending Senate legislation."

Half the text of Kennedy's talk to the medical writers was devoted to the recombinant DNA issue. In that half he mentioned only one name—Cohen.

"The new work of Dr. Cohen at Stanford raises serious questions as to whether recombinant DNA can ever produce a 'novel' organism," Kennedy said. "Dr. Cohen believes that by using this [gene-splicing] technique, scientists can only duplicate what nature can already do."

What curious wording for a man of Kennedy's normally sharp perceptions!

Didn't he realize that the discoverers of recombinant DNA technology had never done anything that was truly original with them—that they were only mimicking nature?

Didn't he know that Peter Lobban, in describing a method of gene splicing for the first time in 1969, had simply told how he and others had observed the symbiotic behavior of bacteria and viruses under certain circumstances and had then proposed to make those circumstances conveniently available to encourage the microorganisms to do what came naturally?

Had Cohen, in performing his own pioneering experiments with plasmids, ever claimed that the work the plasmids and their bacterial hosts had done for him was work he had taught them to do?

What was this "new work" of Cohen's, anyhow, that a United States Senator would so deferentially draw back from it?

I couldn't find out from Kennedy's office, so I telephoned Cohen and asked for a copy of his scientific findings. He told me his report hadn't been published yet, and that tradition had bound him not to disseminate copies until publication did occur. When would that be? In November, he said. November then was three months distant. How did Kennedy know about it if it wasn't published and if scientific tradition banned its circulation until it was published? "My old friend Larry Horowitz had known of this work while it was in progress, and he had expressed an interest in it," Cohen said. Horowitz is a senior member of the staff of the Kennedy subcommittee.

Cohen had been locally notorious as a stickler for observance of ethical niceties by others in the past; how, I asked, did he justify his own clandestine circulation of a supposedly scientific paper for political purposes? Cohen denied having any political motive, but did concede that his paper "had policy implications." When I asked whether he thought it appropriate for a scientist to use an old friend in this way, Cohen contradicted his earlier statement: Horowitz really wasn't an old friend—just an acquaintance, an M.D. who was on leave from his Senate job in 1977 and occupying an office across the hall from Cohen's office while completing a term as resident physician at the Stanford University Hospital. Besides, Cohen said, the paper he gave to Horowitz had, in some way he didn't understand, got "blown out of proportion." But he stuck to his determination not to circulate the paper even to fellow scientists competent to judge its value.

Bureaucratic Washington being a notoriously leaky place, bootleg copies of the Cohen paper inevitably began surfacing here and there. Victor Cohn, of the *Washington Post,* wrote an item in which he advanced the interpretation that any recombinant DNA event that scientists can bring about is commonplace in nature. What Cohen actually wrote in a "Dear Don" letter to NIH Director Fredrickson (bootleg copies of which also appeared in almost magical fashion) was that the unpublished Cohen data provided "compelling evidence" to support the proposition that recombinant DNA molecules constructed in the test tube with the help of the EcoRI enzyme "simply represent selected instances of a process that occurs by natural means."

Cohen declined an invitation to testify about his experiment at the November hearing called by Senator Stevenson. The reason Cohen gave was the now familiar one: the rules of science did not allow him to discuss the subject of his paper until the paper was published. Actually, Cohen had an appointment on that hearing day to discuss his paper in private with Congressman Paul Rogers, chairman of the House Commerce Subcommittee on Health and the Environment.

Cohen's absence from the Stevenson hearing room did not prevent discussion of his paper for the open record, however. Another witness Stevenson had summoned was Jonathan King, who said in public testimony what was being said privately by other scientists who either had picked up rumors of the Cohen paper or seen smuggled copies of the data. Cohen had observed that *E. coli* K12 behaved in a certain way at a frequency of 10^{-9} (1/1,000,000,000), or one in a billion. The conditions under which that behavior occurred were laboratory conditions. So far as anyone knew, they did not prevail in nature. King testified:

> I see no relationship between Cohen's claims and the results [that will be] published. These show that by the use of very potent recombinant DNA technologies, one can generate certain rare events in bacteria in the lab. To conclude, from the fact that [genetic recombination between unrelated organisms] could be generated in the lab, that genetic recombination between unrelated organisms has been going on in nature seems to me totally unfounded. It is equivalent to saying that since we can construct typewriters in factories, typewriters must assemble themselves spontaneously in nature.

King deplored the "abrogation of the normal process of scientific decision making." "In place of published data, open to all for examination and perusal, we have data by personal letter and private phone call," he said. "I myself have been forced into this by exigencies of the situation. But it is not the way to arrive at enlightened public or scientific policy. It is the enemy of both the scientific and democratic endeavor."

King blamed the departure from established scientific norms on "commercial and pseudoscientific pressure to proceed rapidly" and on the "narrow base and representation of those making policy on recombinant DNA research." The underlying danger in gene splicing, he reminded, was the possibility of transfer of DNA from hybrids deliberately constructed in the laboratory to "wild strains, of, for example, bacteria . . . which, in general . . . it will not be possible to rid our environment of . . . , for we have no cleanup mechanisms." "Those sectors of the scientific community with the most relevant experience—public health, pollution, microbiology, microbial ecology, wildlife diseases, occupational health—have essentially been excluded from the proceedings," he asserted. "In essence, a very narrow sector of the research community has maintained control over decision making and policy. Primarily . . . these people represent . . . those interested in the application of the technology and related technologies. . . . The confidence of many of us in the NIH's handling of the issue . . . has [been] undermined."

The issue of *Proceedings of the National Academy of Sciences* that was to carry Cohen's report had still not been circulated when Dr. Philip Handler, president of the academy, appeared before the Stevenson committee to testify. He said he found it "something of an embarrassment that Congress should be entertaining legislation designed to protect the American people against alleged extraordinary risks, the nature and magnitude of which remain a matter of speculation." Some of his auditors in the hearing room were much more embarrassed by his willingness to follow and thereby by implication endorse Cohen's precedent in injecting an unpublished experiment into a political debate.

"Stanley Cohen," Handler said, had "demonstrated that when bits of mouse DNA were taken up by cells of *Escherichia coli,* they were trimmed to size and built into an already existent plasmid within the living host cell by exactly the same kind of enzymatic 'cut and paste' procedure that scientists have been using to introduce a foreign DNA into plasmid DNA before putting the plasmid back into *Escherichia coli.*" "Thus," Handler said, the "laboratory technique [of] using enzymes to make 'recombinant DNA' appears to be nature's own technique for accomplishing the same kind of chemical gene transfer."

Had the president of the academy, too, forgotten that it was nature who showed the scientists that it was possible to perform certain tricks? Had the whole scientific community altogether lost its sense of humility?

Although he took care not to puncture Cohen's exciting balloon, the academy head did qualify his testimony with a sentence that wary legislators (who had the wit to do so) could identify for themselves as a semantic escape hatch for both Cohen and himself to use later if the much touted experiment should not be confirmed. That sentence read: "The full extent of such spontaneous transfer of genetic information among species in nature remains to be ascertained."

After disclosing for the first time that (1) he had recently appointed a special Academy Panel on Risks and Benefits of Recombinant DNA Research, (2) the panel had been chaired by Dr. Maclyn McCarty of Rockefeller University, one of the three men who originally identified DNA as the stuff of genes, and (3) the panel had turned in its findings only the day before Handler's testimony, the academy president pleaded that scientific inquiry be given the protection of the First Amendment to the Constitution. He expressed the fervent hope that Congress would accept the academy panel's conclusion that the NIH guidelines were enough to guard the public safety. "No more than the bare minimal arrangements that would suffice to maximize compliance" with the guidelines, he insisted, should be provided for in any legislation.

Then the academy chief, who presumably is to be taken as a model of appropriate scientific behavior, testified about yet another experiment that had not been published. This one had not even been accepted for publication. "Herbert Boyer and his colleagues," Handler announced, "have just introduced into *E. coli* K12 all the genetic information required to synthesize somatostatin, and the bacteria merrily engage in its production."

What had been introduced into the *E. coli* K12 was a synthetic gene, and what the *E. coli* K12 had reproduced was somatotropin, the chemical precursor of somatostatin, from which the somatostatin could be removed by chemical means after the bacteria that had produced the precursor were killed. It was the closest that anyone had yet come to replicating a gene of a higher eukaryote in a prokaryote and having the gene express itself by making a protein. Somatostatin is a protein. It consists of only fourteen amino acids, It is a human hormone, first discovered in the hypothalamus and identified as the messenger that inhibits the anterior pituitary gland from releasing the growth hormone. Somatostatin has also been found since at certain places in the human gut and in the brain.

Handler's unorthodox announcement of this "scientific triumph of the first order" drew an immediate protest from one of the most respected science writers in the country, David Perlman, of the *San Francisco Chronicle.* In a letter published in *Science* (November 25, 1977), Perlman began with a reminder that the "relationships between the scientific community and the press have often been uneasy. Many scientists are convinced that reporters are frequently inaccurate, that they oversimplify, and that in their rush for headlines and deadlines they fail to wait for completion of the orderly processes of scientific journal review and publication." Perlman said:

> My practice has generally been to await publication in refereed journals, or public discussion at scientific meetings, before leaping to my typewriter with news of the latest breakthrough. I was dismayed, therefore, to note the revelation of Herbert Boyer's most recent recombinant DNA work by Dr. Philip Handler, president of the National Academy of Sciences, in testimony before a Senate subcommittee on 2 November. At the time of Dr. Handler's disclosure that somatostatin had been produced in hybrid bacteria, using recombinant techniques, the manuscript by Boyer *et al.* had not yet been accepted for publication by *Science.*
>
> A double standard of scientific announcement seems to be operating here: the "orderly processes" of refereeing and publication remain in force for journalists and the public. But when the political process is operating in Congress—in this case, apparently, the spectre of political regulation for a new field of science—then the rules of science go by the board, and the public learns of a new scientific triumph via a congressional hearing rather than through the pages of *Science* or the annual meetings of the American Society of Biological Chemists.
>
> I do not believe that a political debate is an appropriate forum for scientific announcements—especially when the scientific community itself purports to guard its traditional processes so zealously. The propriety of Dr. Handler's testimony, however politically useful, should, I believe, be widely discussed.

The propriety of Handler's testimony certainly should be widely discussed, and for more than one reason. Apart from flouting the traditional rules of scientific communication, the president of the academy testified at such a time and in such a way that his statements could be and were interpreted by some as an attempt to circumvent any punishment that might be under consideration for the perpetrators of a much more serious breach of scientific ethics—namely, the violation of the very NIH guidelines that Handler was insisting needed no more than minimum legal sanction to assure their observance.

The transgression of the guidelines occurred in the same P3 laboratory Janet Hopson had written about in the *Smithsonian* magazine. A somewhat disjointed news report of the event appeared in the September 30 issue of *Science* under the heading: "Recombinant DNA: NIH Rules Broken in Insulin Gene Project." Someone working in the laboratories of the Department of Biochemistry and Biophysics at the University of California, San Francisco, had used an unapproved plasmid in a P3 experiment. The modus operandi was virtually identical to the supposedly hypothetical MO that Hopson had attributed to two San Francisco researchers in her *Smithsonian* article three months before: Step 1—disregard the rules in a clandestine experiment; Step 2—keep the results secret; Step 3—repeat the experiment when suitable circumstances arise. The same scientist who had charged Hopson and the *Smithsonian* with irresponsible dissemination of false information was chairman of the Biosafety Committee responsible for thwarting evasions of the rules in the San Francisco P3 lab at the time the rule violation occurred.

Just how *Science* got wind of the violation was not explained in that journal's news report of the incident. But the first clue apparently came from a sharp-eyed reader of the announcement, in *Science*'s June 17, 1977, issue (in subscribers' hands on June 10), of successful completion at San Francisco of an experiment in which the gene that codes for production of insulin in rats had been replicated in *E. coli* K12 with the help of plasmid pMB9, named for its creator, Mary Betlach, a worker in Boyer's lab. The insulin coding gene had been reproduced, but it had not expressed itself by making the insulin protein.

Whoever the perceptive reader was, he was also scientifically knowledgeable enough to know that pMB9 had not been approved by NIH as a microorganism safe for use in recombinant DNA experiments of the P3 class until April 18. When the lag between submission of the report of the experiment to *Science* and *Science*'s acceptance and publication of the report was allowed for, only a few weeks were left for performance of the experiment. The logical question arose: Was that experiment actually done in the manner described in *Science?* Could the experiment possibly have been done in such a short span of time?

When members of the San Francisco Biochemistry and Biophysics Department were asked the question, one said, "It is conceivably possible to do such an experiment in three weeks if everything works perfectly the first time, but you know as well as I do that science never works as well as you hope." Another said, "Well, what can I say? It did work well. We were all set to go." *Science* carried those two quotations in tandem, followed by the cryptic observation: "Members of the [experimental] team concede that an earlier experiment took place but say it was aborted halfway through and before any pertinent information had been gained."

It turned out that the earlier experiment had been performed sometime early in 1977 and had involved use of a plasmid more versatile than pMB9—that is, pBR322, named for the two men in Boyer's lab who together created it: Francisco Bolivar and Ray Rodriguez (who had been Hopson's instructor during her stay in the lab). NIH did not approve pBR322 for use in recombinant DNA experiments at the P3 level until July 7.

By the time the *Science* news staff got around to probing into the matter, the San Francisco Biosafety Committee Chairman, microbiologist Dr. David Martin, and Dr. Howard Goodman, chief of the laboratory where insertion of the plasmids into *E. coli* K12 took place, both were out of the country. It was not unusual for recombinant DNA research scientists to travel thousands of miles and back to argue in support of their work, but no one suggested that Martin and Goodman return home to clear up this embarrassment. On the available evidence, *Science* could only accept the scientists' protestations of innocence. The pBR322 plasmid was approved later, so it could not be held to be dangerous. The only criticism that could be documented was that the minutes of the San Francisco Biosafety Committee meetings, which were open to public inspection, did not give a full accounting of the affair that would be comprehensible to the lay public.

Senator Stevenson trustfully accepted the appearance of innocence. But a canny member of his legislative research ménage could not stop wondering about a couple of sentences in the *Science* news report: "The [rat insulin gene experiment] team had worked in unusual secrecy, which many [in the laboratory] regarded as inappropriate in an academic setting as well as disruptive. 'People would stop talking when you came into the room, or change the subject if you tried to make conversation about how the insulin project was going,' says UCSF biochemist Brian M. McCarthy." McCarthy was a member of the UCSF Biosafety Committee. If he had been curious about the reasons for the secrecy, maybe Senator Stevenson ought to be curious too.

The first witness to appear before the Stevenson subcommittee in regard to the NIH guideline violation was NIH Director Fredrickson. In an introductory statement he expressed more interest in removing regulation of recombinant DNA experiments as quickly as possible than in enforcing the rules that were then in effect. He accepted the UCSF scientists protestations of innocence, hinted that passage of the Kennedy bill would cause more such episodes, and urged instead that "maximum governance" of experiments be kept "at the level of the institution where the research takes place." Stevenson said he hoped everyone would understand his respect for the importance of the work being done at the UCSF lab but then went straight to the point:

> *Stevenson:* NIH guidelines have been violated. Is NIH proposing to withhold funds from [the] university as a result?
>
> *Fredrickson:* No. . . . On the basis of the information available to us we have come to the conclusion that an honest mistake was made by the investigators and by the institutional Biohazards Committee, particularly by the investigators, partly through a system of communications which we have since improved. . . . I think at that time [the communications] were faulty enough to allow them to have concluded that the vector [pBR322] was certified when, in fact, the process had not been completed.

Fredrickson was obliquely referring to the shoestring basis on which Dr. William Gartland was trying to oversee the recombinant DNA research, for a long time without even a secretary of his own, and then with a secretary and one assistant. Only a quarter of a million dollars was spent in the first year of the Office of Recombinant DNA Activities, and it was spent against the current of history, for NIH looked upon itself as a fairy godmother with open moneybags and a disdain for the idea that any scientist should ever need a policeman to keep him from dipping into the till for experiments the public might look askance at.

> *Stevenson:* So it is your opinion that if the mistake which produces the violation of NIH guidelines is honest then funding should not be withdrawn or perhaps [should be] continued [so] that the institution in question should also be permitted to get the patent, to exploit the fruits of that research for its own commercial gain?
>
> *Fredrickson:* I would not like to extrapolate this particular decision . . . to all such, Mr. Chairman. But I do believe that this institution should not be penalized. . . . They destroyed this clone as soon as they were aware that they were outside the guidelines. . . .
>
> *Stevenson:* Does such a policy encourage compliance with the guidelines or does it encourage innocent mistakes?
>
> *Fredrickson:* In this instance, Mr. Chairman, I would look at the other side. . . . I am encouraged rather than discouraged by this particular incident because I think it shows that peer pressure and an

open system probably can be more effective in policing itself than one hidden and requiring police activities for disclosure.

Stevenson: Just how did NIH find out about this mistake?

Fredrickson: NIH first became aware . . . at the time that a reporter called . . . regarding an article . . . for *Science.*

Stevenson: *Science* magazine?

Fredrickson: That is correct.

Stevenson: You said earlier that it [the violation] had been reported. Now you indicate that NIH found out about it by reading a magazine. Is that how this system is supposed to work?

Fredrickson: No. It is not supposed to work that way, Mr. Chairman. We should have been informed immediately by—as soon as possible by the institutional Biohazard Committee.

At that point, Senator Harrison Schmitt, a former NASA astronaut and now a member of the Stevenson subcommittee, asked the chairman's permission to interrupt to clear up the sequence of events. The questioning proceeded.

Schmitt: I want to make sure that the record is clear. . . . The reporter preparing the article called NIH for comment; is that correct?

Fredrickson: Yes.

Schmitt: So, prior to publication, you became aware of it. Then, when you contacted the Biohazard Committee at the university, were they aware of the situation?

Fredrickson: Gartland wrote to the chairman of the Biosafety Committee on October 11. He got a reply on October 25.

By the time of that exchange of letters, Dr. Martin had gone to England and the chairmanship of the UCSF Biosafety Committee had been taken over by Dr. James Cleaver. Cleaver made his own investigation as a result of Gartland's query.

Stevenson: The October report from Dr. Cleaver says that Dr. Boyer informed members of the insulin research team on February 4 that pBR322 had not been certified, yet work continued on the clones obtained [from pBR322] until March 3. Are you undisturbed by those circumstances?

Fredrickson: I am disturbed, Mr. Chairman, that work of any kind proceeded.

Stevenson: Isn't it true that the report also states that the Biosafety Committee was informed . . . [of the use of pBR322] on June 3, three months after the experiments had been halted?

Fredrickson: That is correct.

Stevenson: Is that disturbing?

Fredrickson: Yes, it is disturbing.

Stevenson: That being the case, do you plan to take any action?

Fredrickson: Well, I am sure that NIH in its revision [of the guidelines] must be more explicit and more demanding of all its institutional review boards and also of our communications with them.

Stevenson: But Dr. Fredrickson, in the instance I am referring to now, there wasn't any confusion. They knew that the plasmid had not been certified. Dr. Boyer informed members of the team on February 4. Your communications, it would appear, could be improved. But in the situation I am referring to now, there wasn't a failure of communications. There was a failure to respond to the fact. What do you do about that? It is not just the lack of authority at NIH that is disturbing. It is the lack of authority for any acceptable chain of command within the institution that is disturbing.

Fredrickson: Yes.

Stevenson: I can't make out . . . who is responsible.

Fredrickson: Well, the Biosafety Committee chairman is still an agent of the institution. The ultimately responsible figure must be as determined by the institution, either a Dean or the Chancellor of that campus.

Stevenson: Well, that is pretty ultimate. That is a little like saying the President of the United States is responsible for NIH.

Dr. Fredrickson, one of the messages I get from what you have said this morning and been unable to say in some instances is that it is one thing for NIH to lay down the guidelines for research, but another to enforce them. You would prefer not to have the enforcement responsibility at NIH; is that correct?

Fredrickson: Yes, sir.

The next witness was Dr. William Rutter, chairman of the UCSF Department of Biochemistry and Biophysics. He read a prepared statement in which he admitted that pBR322 had been used in the rat insulin gene experiment early in 1977. It was used, he said, because he had been unofficially informed in January that pBR322 had been approved by NIH. In March, however, he said he learned that the process of approval had gone only as far as the NIH Recombinant DNA Molecule Program Advisory Committee, and that a final okay by the NIH director was lacking. "Immediately," Rutter testified, "we . . . made the decision to stop the . . . experiment." Continuation of the work with an earlier approved plasmid, pCR1, was attempted without success. Then pMB9 was approved, and it was used successfully. The results achieved with pMB9 "were the ones reported in June in *Science*." All the experiments were carried out under P3 containment conditions. The host bacterium for both pBR322 and pMB9 was Curtiss's "disarmed" χ1776 strain of *E. coli.*

If everything happened as innocently as Rutter claimed, why did the UCSF team withhold information from NIH about the misuse of pBR322? Rutter asked the question rhetorically and answered it with a string of technicalities that put all the blame on NIH. The agency's procedures were loose and informal, he said; he and his colleagues were confused by conflicting information. The guidelines themselves, he said, were open to various interpretations at a number of points. Altogether, he made a plausible case until he reached the final point in his rationale, which had been carefully composed in advance and went like this:

> The arguments against informing NIH were associated with the inflamed social and political climate that existed with respect to recombinant DNA technology. The situation . . . was unprecedented:
>
> Scientists, of their own volition, had agreed first to a moratorium, and then to a cautious experimental demeanor using these revolutionary new approaches. The NIH Guidelines were within that spirit.
>
> The debate about recombinant DNA rapidly entered the public domain. This issue appeared to coalesce various concerns regarding research on atomic energy, research on biological warfare, environmental pollution, etc. The spectre of a biological holocaust was portrayed. This issue was used by some, including scientists, as a vehicle for their own political or philosophical motives. The press, among others, has sometimes fanned the flames of controversy. Cloning molecules has been linked to cloning human beings. Imaginary, and sometimes sensational, scenarios have been depicted. This combination of science and fiction makes good copy. Consequently, the value of scientific exploration has been questioned. Scientists have sometimes been depicted as sinister instead of useful members of society. Repressive and punitive legislation was being considered.
>
> We felt that informing NIH would inevitably lead to public

disclosure of this incident, and that would exacerbate the situation. It would be unfair to NIH and the Recombinant DNA Committee who were scurrying to regularize their procedures and to scientists who were trying to accommodate to being regulated. In the end we chose not to inform NIH formally, but to take the most conservative approach to the experiments themselves. We destroyed the clones.

There was no question of a cover-up; more than 25 people in the department knew directly of the experiments. I have never asked a single individual not to discuss this matter. There was no doubt that the incident would be broadly known by scientists. On the other hand, we did not publicize that incident, nor raise the issue in public. I do not question the right of the public to be involved broadly in decisions regarding scientific inquiry. However, it would seem a prerequisite that they be well and impartially informed.

Rutter was seeking a curious kind of impartiality. He had not hesitated to participate in the pro-recombinant DNA lobby foray into the Congressional offices from the headquarters of AAAS on August 1. Goodman had been a leader of the group that drafted the letter to Congress from the Gordon Research Conference on Nucleic Acids in June. Boyer had spent many days of the spring and summer in Washington, conducting his own personal lobby against the Kennedy bill, interspersing professions of fear of "persecution" with threats to quit research or even leave the country if regulations he didn't approve were imposed on his experiments. Stevenson may not have been acquainted with those activities. In any case, he did not ask Rutter to define impartiality. That was unfortunate. The record would have benefited from a clear statement of it before the questioning began.

Stevenson: Dr. Rutter, you stated that in January you learned unofficially that this plasmid [pBR322] had been approved.

Rutter: That's correct.

Stevenson: How did you learn that it had been approved?

Rutter: I learned directly from a conversation with Axel Ulrich, an individual in the laboratory, who in turn learned from a person in Herb Boyer's laboratory, who in turn had been called by a member of the Recombinant DNA Committee [of NIH] stating that the Committee had approved pBR322. I said, "Yippee!"

Stevenson: Who was the person on the [NIH] Committee?

Rutter: Dr. Betty Kutter.

Boyer, sitting beside Rutter at the witness table, interrupted at this point.

Boyer: I can perhaps provide some reconstruction of the chronological events. We were concerned about developing more versatile and more efficiently contained plasmids and had been working on this for some time. Around the end of 1975, we were trying to find out exactly how certification could be obtained for these plasmids. I contacted members of the DNA Advisory Committee and also talked to Dr. Gartland at NIH. They recommended that I follow a protocol that was used for a plasmid that had been approved earlier. That is what we did, and the data so obtained was submitted to NIH for its consideration and approval.

In January [1977], we were told that tentative approval had been given by the Committee for both plasmids pending receipt in writing of the information from Dr. Falkow's lab.

Stevenson: When was that?

Boyer: January of '77.

Stevenson: So you both learned by this somewhat circuitous route of what? What was it Dr. Kutter was reported to have said?

Boyer: The exact wording was that the plasmids were given tentative approval pending receipt of the information from Dr. Falkow.

Stevenson: Dr. Rutter, is that what you had?

Rutter: That nuance had not reached me.

Stevenson: Are you suggesting that that is a nuance?

Rutter: Well—

Stevenson: "Tentatively approved subject to the receipt of additional data." Is that a nuance? That is accurate. That is what in fact had happened.

Rutter: Senator Stevenson, the word that we received from Dr. Kutter, as far as I know—and I had a conversation with her over the telephone—was that the Committee had approved pBR322 as a plasmid.

Stevenson: But Dr. Boyer said that he heard from her, indirectly in January, that the Committee tentatively approved, subject to the receipt of additional data; did you not?

Boyer: This is true. You must remember that I am trying to recall all of this in retrospect. I didn't write it down.

Stevenson: Isn't that exactly the point? Would reasonable men, let alone scientists, have proceeded without some confirmation, without something in writing, without something that wouldn't put you in this preposterous position today? You say you don't want legislation. If there is legislation, you gentlemen would be the authors of it.

Boyer: Senator, I would like to point out that the use of the other plasmids that had been certified was never made known to us in writing. It was all by word of mouth.

Stevenson: So one act of negligence deserves another?

Now, let's turn from the beginning to the end. Dr. Rutter, you stated that you learned at a meeting in early March that pBR322 had not in fact been certified by the director of NIH. You immediately made a decision to stop the cloning experiments. The memoranda dated October 14, 1977 from you and Howard M. Goodman to Dr. Cleaver, who was then chairman of the Biosafety Committee, says that on or around February 4, 1977, Drs. B. J. McCarty and H. W. Boyer called an ad hoc meeting with the members of Dr. Goodman's laboratory to tell them that a logbook for the P3 facility was being instituted. Verbal instructions for use of the logbook were given and a written example was provided on the first page of the book. And Boyer informed them that pBR322 was not yet a certified vector.

My first question is to you, Dr. Boyer. Did you, as Dr. Rutter's memorandum states, know—and on February 4 inform others—that pBR322 was not yet certified?

Boyer: At that meeting it was stated that the pBR322 plasmid, although approved, had not been officially sanctioned by NIH in a written statement.

Stevenson: So you did know at that point that it had not been certified; is that correct?

Boyer: I personally knew, yes.

Stevenson: How did you come by that knowledge?

Boyer: I had been in constant contact with Bill Gartland about this.

Stevenson: You said earlier that on January 16 you were told by Dr. Gartland that it had been approved subject to additional data. On February 4, you are stating to others that it had not been certified. . . . Did you know on January 16 that it had not been certified?

Boyer: I can't say for sure that I really understood the difference between certification and approval on January 16.

Stevenson: At some point you became aware, did you not, that it had not, in fact, been approved by the committee?

Boyer: Yes.

Stevenson: When was that?

Boyer: After the Miami meeting, several weeks after.

Stevenson: And before February 4, presumably?

Boyer: Presumably.

Stevenson: Dr. Rutter, your testimony states that "we" learned at a meeting in early March that pBR322 had not, in fact, been certified by the director of NIH. Who, Dr. Rutter, does "we," the plural, refer to?

Rutter: "We" refers to Dr. Howard Goodman and myself. Dr. Goodman was on sabbatical last year, in Japan. He returned about the middle of February, and shortly thereafter went to a meeting at which recombinant DNA work was discussed peripherally, among other things. A meeting in Utah—Park City, Utah. I was speaking at another meeting in Houston, Texas. According to Goodman, Dr. Gartland's answers to questions put to him from the floor [of the meeting in Utah] were the first indication Goodman had that pBR322 was not certified. He immediately tried to call me, I think, on February 3rd. He did not reach me that day but did reach me in Houston on February 4th. I was surprised and perplexed by his information and agreed that we should immediately cease experiments.

Stevenson: When did you learn that Dr. Boyer and others knew that the vector [pBR322] had not been certified? When did you find out that they knew as of February 4?

Rutter: It was sometime in March that I learned that Herb had known and that Brian McCarty had known. We are all colleagues in the same department. We have meetings together. We are friends. Nevertheless, this information [about pBR322's status] had not passed between us. I did not know of this matter until March 4th, as I told you.

As a former spaceman, Senator Schmitt was naturally sympathetic with the problems of scientists who are working under competitive pressure, as the San Francisco insulin group had been. Throughout the Stevenson hearings, he asked questions reflecting that point of view. But the answers Stevenson was getting began to irritate Schmitt and he asked the chairman's leave to interrupt.

Schmitt: Dr. Rutter, are you saying you were not at the February 4 meeting?

Rutter: I was absolutely not at the February 4 meeting.

Schmitt: Was Dr. Goodman there?

Rutter: He was in Japan.

Schmitt: Other members in the laboratory obtained this information from Dr. Boyer that pBR322 was not yet a certified vector?

Rutter: That is correct.

Schmitt: Those other members of the laboratory then apparently did not think this was significant enough information to transmit either to you or to Dr. Goodman; is that correct?

Rutter: I won't say they didn't think it was important. What I do say is they did not transmit that information to me or Dr. Goodman.

Schmitt: Are these people graduate students?

Rutter: They are postdoctoral students.

Schmitt: Had they been involved in the research for some years?

Rutter: For at least a year.

Schmitt: Would you say they were conversant with the NIH guidelines as they stood at that time?

Rutter: Yes, I would.

Schmitt: That concerns me, Mr. Chairman. I guess my main feeling is that the laboratory procedures were not very systematic, were not really what are wanted in a research lab.

Rutter: That is exactly right.

Schmitt: In fact, a logbook appearing on February 4 is tremendously disconcerting to me. Not only that you were operating under guidelines, but under the P3 portion of those guidelines. You were in an area where keeping records can be of tremendous importance when you go back and try to understand some particular result that was unexpected or even a result that was expected. I think you certainly deserve some criticism—or your lab does, whoever is responsible for it—for not carrying on more responsible scientific research.

Stevenson resumed his questioning by asking for confirmation of February 4 as the date when the logbook was instituted in the P3 laboratory.

Boyer: Yes. That is correct.

Rutter: Although the logbook was instituted on the basis of a conversation between Herb Boyer and Brian McCarty and individuals in the two laboratories, the book was not seriously employed until March. To the best of our knowledge, dates in March were the first entries in that book after the explanatory entry made by Herbert Boyer.

Stevenson: Wasn't the use of . . . pCR1 . . . recorded retroactively to February 1?

Rutter: It was recorded retroactively, to the best of our knowledge, from March 11.

Stevenson: It was recorded on March 11?

Rutter: Yes. To indicate that the experiments had been started on February 1, a retroactive entry was made on March 11.

Stevenson: And was the use of pMB9 also recorded?

Rutter: That research was recorded in the logbook when it occurred.

Stevenson: Was the use of the pBR322 plasmid, the one in question here, also recorded?

Rutter: It was not.

Stevenson: Why was it not recorded even retroactively?

Rutter: I can't answer that.

Stevenson: Dr. Boyer, can you explain why the pBR322 plasmid, unlike the others, was not recorded even retroactively?

Boyer: No. I had no knowledge that the plasmid was being used in those experiments. I have no explanation for that.

Stevenson: You had no knowledge that that plasmid was being used at that time?

Boyer: I was not aware of the results of those experiments, or the fact that they were ongoing, until the meeting in Utah.

Stevenson: Who does know? Who is in charge? Who was at the time?

Boyer: These experiments were not carried out in my laboratory.

Stevenson: Who was in charge of the laboratory in which they were being carried out?

Rutter: The laboratories in which the cloning experiments actually were carried out is Dr. Goodman's laboratory. He was in Japan on a sabbatical. Various aspects of the work going on in his laboratory were under nominal supervision of myself as the head of the department and as a collaborator on this project, but also were shared by Drs. McCarty and Boyer.

Stevenson: You said that when the fact that the plasmid had not been certified was learned on March 3, clones were destroyed?

Rutter: The experiments ceased at that time for a period of a week or so. The clones were not destroyed while we were going through the process of trying to figure out what was going on and what should we do immediately.

Stevenson: A week or so? Can you be more precise?

Rutter: I do not know the precise date.

Stevenson: During that period of time were you contacting NIH or someone to find out whether and when the plasmids would be certified?

Rutter: Yes, I was. Dr. Gartland said that there was no question of certification, that the plasmids had been approved by the Committee, and that formal approval by Dr. Fredrickson was imminent. He suggested within a week or 10 days at that time.

Stevenson: Was approval in fact obtained within a week or 10 days?

Rutter: No. It was not. Formal certification did not occur until July 7th.

Stevenson: Was that why you waited a week or 10 days to destroy the clones?

Rutter: We were not sure exactly which action to take. When it was clear that there could be delays in certification, we chose not to inform NIH formally, but to . . . destroy the clones.

Stevenson had been vexed for some time. He hadn't anticipated anything like this when he started his hearings. Schmitt's annoyance was growing too. It became apparent when Stevenson turned the witnesses over to him.

> *Schmitt:* Dr. Rutter, you say there was no cover-up; but you did make a conscious decision not to inform the NIH. . . . I think . . . the rationale that some knowledge in the hands of the public would be dangerous or counterproductive is not really an acceptable rationale, even though I will admit that sometimes a little information can cause great misunderstandings. It is also inconsistent for such a rationale to be professed by people who argue against the idea that there is some knowledge science should not seek.
>
> One reason I am in public office is that I think the public needs to be better informed on a lot of issues. But making an independent decision on what the public should know and what the public should not know is, I think, counterproductive. Maybe your experience is starting to prove that.

Schmitt then gave the witnesses an opening to say that the pressure of competition had clouded their thinking about public disclosure. But Rutter insisted that the competition "did not dominate our demeanor." After Stevenson adjourned the hearing for the day, a half dozen federal government officials, all involved in one way or another with the granting of funds for science, gathered in the privacy of a men's room down the hall from the hearing chamber. "It isn't often that you hear a United States senator call a prominent scientist a liar to his face in public," one said. "I just can't believe what I heard in there," said another. "It's bound to hurt science in the long run. It may even have a short run effect on budgets for science."

The hearing continued on the following day with testimony from several panels of witnesses. The answers Stevenson got from the members of one panel had extraordinary significance because two of them had been prominent participants in that phase of the science lobby's activities that was directed at Senator Nelson and which resulted in the writing of his amendments to the Kennedy bill. The lobby's strategy had called for the enlisting, wherever possible, of a reputable scientist from the home district of each key legislator. In Nelson's case the local boy who made good was Dr. Oliver Smithies, who spoke for the University of Wisconsin, the Genetics Society of America, and the Federation of American Societies for Experimental Biology. The second member of that panel was Halvorson, the ASM president, who did what few scientists have the humility to do—asked the advice of people experienced in gauging the reactions of the lay public and then (an even rarer occurrence) followed the advice. Others on that same panel were Dr. Joseph Keyes, speaking for the Association of American Medical Colleges; and Dr. Frank Young, professor and chairman of the Department of Microbiology at the University of Rochester and chief microbiologist at the Strong Memorial Hospital. The main topic of discussion was appropriate procedures for licensing recombinant DNA research. The panel agreed that there should be licensure, that license holders should be monitored by local institutional biohazard committees, and that the committees in turn should be monitored by some federal government agency, preferably the Center for Disease Control at Atlanta.

After inviting and receiving formal statements from the panelists, Stevenson opened the questioning.

> *Stevenson:* Let's say the Center for Disease Control is the national monitor and it discovers that the institutional monitor, a biohazards committee, has failed to monitor effectively or had monitored and discovered a violation but failed to take appropriate remedial steps. At that point, what does it [CDC] do?
>
> *Halvorson:* At that point the institutional license should be revoked.
>
> *Stevenson:* The institutional license? What do you mean by the institutional license?
>
> *Halvorson:* Well, we earlier addressed the question of the biohazard committee within an institution monitoring its own individual laboratories. I think that is the easiest and best way of getting peer pressure and control over the experiments within a given institution. Should the institution fail to meet that responsibility, then is the appropriate time for that institution to lose its license.
>
> *Stevenson:* In effect, to get shut down?
>
> *Young:* That's right. The loss of license is the gravest threat one could impose.
>
> *Stevenson:* Now, having laid out this hypothetical situation, let me put a case to you. In January, a facility conducting federally supported recombinant DNA research—giving it the benefit of a doubt—understands that a host-vector has been approved by NIH. Early the following month it discovers that is not in fact the case. It continues its research with the noncertified vector. The following month, say about four weeks later, the chairman of the department in question also discovers that—as researchers have known for a month or so—the vector was not certified. A month of unapproved, unauthorized research has taken place. He stops the experiment then, but the clones aren't destroyed for another, oh, week or ten days. The biohazards committee isn't even notified for—well, until sometime in June; and its notification is kind of indirect or casual, at most;

nothing happens. The NIH is never informed. It discovers it from a magazine reporter. Could you, Dr. Smithies, call that series of events a technical violation?

Smithies: I think it is a very deplorable set of circumstances. There are two things that would make me question whether it is technical or not. One question I would ask is whether the host-vector system which was used was eventually certified or not.

Stevenson: After the fact, it was certified.

Smithies: Were the investigators aware that the host-vector system had passed all of the tests through the scientific part of the approval system?

Stevenson: No. They were aware that it had not.

Smithies: That it had not passed them?

Stevenson: That additional data was required.

Smithies: Then I think they had done something that was incorrect. I don't know whether it is appropriate for me to tell you how I would deal with it if I had to.

Stevenson: Well, this is what I am leading up to, how this great system of yours works.

Smithies: I would have to say I regard this as deplorable. I think it is partly technical. I would not yield on that. It is bascially technical. The organism is in fact a safe organism. I would say—

Stevenson: What you seem to be saying is that it is not for the NIH to determine whether it is a safe organism, it is for the researcher. If he is proved right subsequently, he is exonerated.

Smithies: No. I don't think he should be exonerated. I wasn't trying to suggest that at all, sir. I was going to suggest there should be different degrees of penalty. If they had been working with an organism which, there was reason to think, was not safe, then it is not a technical violation. That is a moral and a biological violation. I would revoke the license for that. In this category that you indicate, I would say that they should be shut down for a period of time equal to the time in which they were in violation, as an indication that people can't get away with this sort of situation. There should be some penalty. I think the penalties should recognize whether it was a really serious violation or a marginally serious one, with some gradation. If it were a very serious violation, I would contemplate revoking license for a long period.

Stevenson: Well, in this instance, there was, to say the least, some hesitancy or delay about informing the local biohazards committee. The committee didn't discover it themselves. Then it didn't take any action. Now, if the penalty is to close down the facility, why would the facility close itself down? It wouldn't inform on itself, would it?

Smithies: This is where you have to have a second level of inspection to see whether compliance is being monitored in the correct way locally. You have given a very difficult example. It is an example which those of us in the field are not proud of.

Stevenson: Dr. Young?

Young: I think it is very important that the biohazards committee have a central P3 facility under its aegis.

Halvorson: Could I add a comment to that?

Stevenson: Yes, sir.

Halvorson: We know work is being carried out by people who do not have training in handling hazardous articles. There we have a training obligation. We have urged that the biohazards committees have on them experts knowledgeable in infectious diseases, epidemiology, relevant areas, so in examining other members of the institution who want to carry out the experiments, and monitoring them on a peer basis, they can take into consideration what are the appropriate laboratory techniques.

Stevenson: Any other comments?

Smithies: Yes. I would think that [the DNA program advisory committe of NIH] might consider setting up a regional group of visiting investigators, or visiting scientists. They should really be experts, not just at the technical level. They would go around occasionally, randomly, so as not to have an established pattern, visit a larger institution or an industrial complex and ask to be shown what is being done. I might say that we [at the University of Wisconsin] have a P3 facility that is rather small. We train all our people to assume that they are going to be inspected any time. If you cannot stand inspection on the spot, you are not doing your job right.

Stevenson: Dr. Young, do you agree with that?

Young: I agree with that.

Stevenson: Has the lawyer anything to add?

Keyes: I don't think I have anything to add.

In his testimony, NIH Director Fredrickson had given the Stevenson subcommittee two assurances. One was that the internal NIH communications mess that contributed to the San Francisco caper had been cleaned up. The other was that the annual NIH guideline revisions that had been approved by the Recombinant DNA Molecule Program Advisory Committee of NIH during the summer of 1977 and issued in rough draft form by Fredrickson in September would foolproof the guidelines against all reasonable possibility of violation. How valid were those assurances?

The absence of foundation for the first one became evident while the Stevenson staff was still working on the preliminaries for the report the senator planned to present to the Senate at the second session of the 95th Congress in 1978.

The bad news broke officially at about ten thirty on the night of December 15,1977, the opening day of a two-day meeting of the Advisory Committee to the Director of NIH. Fredrickson had assembled this group of his topmost counselors at NIH headquarters in Bethesda for what amounted to a public hearing on the proposed guideline revisions. Just before the first day's session ended, an hour and a half behind schedule, the director read a letter that Dr. Arthur E. Heming—associate director for program activities of the National Institute of General Medical Services (in which Gartland's Office of Recombinant DNA Activities is housed for budgeting purposes)—had written on the previous day to Henry Meadows, dean of finance and business at Harvard Medical School in Boston. The letter consisted of this one paragraph:

> On December 9, the National Institutes of Health (NIH) received a proposed Memorandum of Understanding and Agreement (MUA) dated November 2, for Research Grant GM 21740-03, Dr. Charles A. Thomas, Jr., Principal Investigator. Based on information developed by an NIH site visit team which visited Harvard Medical School on December 12, NIH is withholding approval of this MUA pending clarification of prior compliance by this investigator with the NIH Guidelines for Research Involving Recombinant DNA Molecules. In the absence of an approved MUA, recombinant DNA experiments supported by NIH funds under Grant GM 21740-03 must not be carried out. You will shortly receive a revised Notice of Grant Award indicating that funds from GM 21740-03 may not be used to conduct recombinant DNA experiments.

According to NIH officials, it was the first time that NIH had even momentarily cut off the funds of any recombinant DNA research grantee on the ground that conditions of the grant had been violated.

It was a particularly shocking case because Dr. Thomas was a member of the original Recombinant DNA Molecule Program Advisory Committee of NIH.

As in the San Francisco violation, ambiguous communication with the NIH and between the NIH and the Biohazards Committee of the local research institution were involved. The situation was much more complex than Heming's letter suggested, and NIH could claim credit for very little initiative in the action that was finally taken.

Especially in the wake of the revelations in San Francisco, the NIH Office of Recombinant DNA Activities should have been alerted by a news item published in the *Harvard Crimson* in November 1977. The news was that Dr. Thomas had quit the Harvard faculty to accept a post at the Scripps Research Foundation in San Diego on January 1, 1978. The *Crimson* attributed Thomas's departure to two causes: first, politics in Harvard's biological chemistry department, which Thomas found distasteful; and second, controversy over Thomas's recombinant DNA experiments.

Thomas was notorious for adamant opinions. He repeatedly walked out of meetings of the NIH Recombinant DNA Molecule Program Advisory Committee to express his contempt for the level of discussion. He was known to have been engaged in recombinant DNA work at the P3 level before the first set of NIH guidelines went into effect. Since that time different people had different ideas about whether he had done experiments at the P3 level in a laboratory cleared by the Harvard Biohazards Committee for P2 work only. Thomas claimed he had never violated the NIH guidelines. That, however, was not the issue addressed by the NIH letter to Harvard.

Under an instruction that Fredrickson had signed on August 26, 1976, every recombinant DNA research grantee was expected to have a Memorandum of Understanding and Agreement (MUA) on file in Gartland's office by November 15, 1976. An MUA was in effect a contract committing the researcher to abide by the NIH guidelines. Before an MUA could get into Gartland's file, it had to be approved by the local Biohazards Committee and signed by an officer of the local research institution who assumed legal responsibility for the institution—no one lower in the hierarchy than a dean; alternatively, a vice-president or the chancellor. From the local institution the MUA would go to the office of the director of research grants at NIH, where it would be scanned and if found acceptable sent on to whichever one of fifty review panels was responsible for appraising the scientific worth of experiments in the particular scientific discipline practiced by the grant applicant. If the appropriate review panel okayed the application for funding, the MUA then would pass to

Gartland's office, where his assistant, Dr. Daphne Kamely, would check it out and transmit it at last to the institute which was to provide the money.

After the announcement of Thomas's imminent departure for Scripps appeared in the *Crimson,* someone who purported to be familiar with his work habits took a complaint to the Environmental Defense Fund in Washington, D. C. It was a timely move. EDF Science Associate Leslie Dach, who had a master's degree in science and several years experience on the staff of the Institute of Society, Ethics and the Life Sciences at Hastings on Hudson, had just been invited to represent EDF at the NIH guidelines' hearing on December 15–16. Dach telephoned Gartland's office early in December and asked if Thomas was operating his laboratory in strict accord with the guidelines. Gartland went to his files and looked for an MUA from Thomas. There was none. Neither Gartland nor Kamely could explain why it was missing, for Thomas had a five-year NIH grant of a half million dollars, and the grant was nearing the end of its third year.

Given Gartland's report of the situation, Dach felt justified in writing a formal letter to the NIH office that handles requests submitted under the Freedom of Information Act. The letter was dated December 6, 1977. Addressed to Ms. Joanne Belk in the NIH Office of Communications, it asked "specifically . . . for all documents relating to work practices and/or compliance with NIH guidelines for research involving recombinant DNA molecules in the Harvard University department of biological chemistry and/or the Harvard University laboratory of Charley Thomas." Dach wrote that he intended to use the answer to his letter in the public interest and would appreciate receiving the answer by December 15. As a matter of convenience, he said he could "pick it up at the NIH campus."

Three days later, an MUA covering Thomas's grant suddenly appeared on Gartland's desk. It was dated November 2, 1977, more than a year past the deadline Fredrickson had set in his memo of August 26, 1976.

Six days after Dach's request for information had come into the NIH offices, three officials of the agency—Gartland, Dr. Bernard Talbot, representing NIH's deputy director for science, and James W. Shriver, representing NIH's "detective" force—went to Harvard for separate interviews with Dean Meadows, Biohazards Committee Chairman Bernard Fields, and Thomas. Thomas told the visitors that he had filed an MUA with the local Biohazards Committee on November 12, 1976, three days before the deadline.

An interesting question was raised by Thomas's claim. Was a local

Biohazards Committee obligated to send NIH an MUA that the committee did not approve?

Other equally if not more tantalizing questions were: If the committee did not approve Thomas's MUA in 1976, why was approval given in 1977? If approval by the local committee was given on November 2, 1977, why did it not reach Gartland's desk until three days after Dach's December demand for the facts?

There was no way for Fredrickson to avoid cutting off the grant at least until the air was cleared. Whatever the date on the MUA, the document still remained unapproved in the prescribed NIH fashion. In practical terms, the freezing of funds did not seem to be more than a rap on the knuckles, for Thomas's departure coincided with the end of the grant year.

For the first time in the history of the recombinant DNA controversy, private industry was involved in a complaint about guideline violation. Miles Laboratories, Inc., an enzyme maker with an excellent reputation, was accused of financing a gene-splicing experiment with a bacterium that had not been certified by NIH. The offending organism was *Bacillus subtilis*—the soil-dweller that Robert Pollack had urged on Paul Berg in the telephone conversation of June 1971. *B. subtilis* does not inhabit the human gut and is not known to be responsible for any human ailments. But it isn't as well known as K12 and many experimenters wouldn't feel comfortable with it. Rochester University Professor Frank Young had been studying it and trying to make it a safe candidate for gene-splicing work. Young's studies were challenged at the NIH public hearing because the NIH guidelines do not cover *B. subtilis.* The complaint proved baseless, however. Young was not doing the *B. subtilis* work with public monies but under private contract with Miles. Young said he had not yet succeeded in breeding the hybrid he wanted. As soon as he did, he added, he would apply for certification by NIH whether or not Congress in the meanwhile acted to regulate privately funded recombinant DNA research.

The remainder of the NIH hearing was taken up with discussion of the guideline revisions recommended by the NIH recombinant DNA Molecule Program Advisory Committee. The proposed changes constituted a package that fitted neatly into the philosophy that had surfaced in secret at the NIH consultants meeting on August 31, 1976, and had flowered in the Gorbach report of the Falmouth workshop of June 20–21, 1977—"everything we have learned tends to diminish our estimate of the risks associated with recombinant DNA in *E. coli* K12." If Fredrickson should endorse the whole package, the gene-splicing lobby would have everything it wanted except possibly the humiliation of former President John F. Kennedy's youngest brother, Edward, in a floor vote in the Senate.

There were five big items in the package, the biggest of all one that would chop at least the bottom third (possibly as much as two-thirds) of the currently active heap of approximately 300 recombinant DNA experiments in this country out from under the NIH guidelines—and thereby out of reach of the law—by redefining the recombinant DNA molecules that NIH wanted to keep under control.

The purpose of the new definition was to recognize an established fact noted in the prologue to this book—namely, that the natural occurrence of certain types of recombinants had been known since 1928. The recombinants that give cause for greatest concern today are the novel ones, those not known to happen naturally but known to be contrivable artificially through use of gene-splicing technology. The proposed amendment to the NIH guidelines therefore specified that recombinant DNA subject to NIH oversight should be only those "molecules that consist of segments of any DNA from different species that are not known to exchange chromosomal DNA by natural physiological processes."

Two other items in the packet of changes proposed for the guidelines would relax restrictions on the most controversial and supposedly risky experiments at the top of the currently active agenda—those involving uninfected cells of adult humans and the cells of mammals other than primates, notably mice, a popular laboratory subject.

Except for embryonic and germ cells, which always have been assumed to be free of infection, human cell lines had been assigned special guards in the original NIH guidelines out of fear that such cells might harbor viruses that cause cancer in humans. In 1977, the NIH Recombinant DNA Molecule Program Advisory Committee had decided that the fears were unduly exaggerated. Millions of dollars had been spent by the federal government in search of viruses that cause human cancers and no such viruses had been found.

When the NIH Recombinant DNA Molecule Program Advisory Committee made its decision in June, and when the first draft of proposed guideline revisions based on that decision were published in September, competent opinion among scientists seemed to be in overwhelming agreement that the proposed guideline change on this point was correctly oriented, although for some unstated reason no one suggested stopping what was indirectly being branded as a fantastic waste of the people's money.

Between September and the mid-December NIH hearing, however, an unexpected dissonance broke the complacency. In November fell the fiftieth birthday of the Fox Chase Cancer Center's research institute at Philadelphia. A party was held to celebrate the occasion. One of the guests was Paul Berg. During a lively exchange of views on the future of cancer research, he advocated vaccination of children against infection by viruses that may cause cancer.

Behind this bit of boat rocking were a series of largely theoretical postulations that few laymen are familiar with. These had been discussed at the first Biohazards Conference at Asilomar in 1973 (see chapter 7) and were published in the proceedings of that conference by the Cold Spring Harbor Laboratory Press. The first established fact from which the theory flows is that cancer cells are present in the bodies of virtually all humans a great deal of the time. Most of these cells are killed off by the body's immunological system before they can gain a threatening hold. The success of the defense is so marked that it has led to the postulate that the immunological system took shape originally very long ago for the sole purpose of fighting off cancer and only later was broadened to include defense against other forms of infection as well.

Finding themselves effectively turned back from those early battles, the theory goes, the cancer-causing viruses adopted a longer-range strategy for their own reproduction, which takes place only in other cells. The strategy called for quiescent entrance into the cells, followed by lodgment in the DNA and there passing along their attributes from one generation to another until some external trigger activates them to cause cancer again.

All this is largely blue sky thinking. To bring it down onto solid earth, Berg fastened on BKU, a virus that is related to the common virus that causes warts. BKU also causes tumors in hamsters and has also been found in the tissues of several human tumors and likewise in some human cancer cell lines. Berg also mentioned another virus called EBV, which is routinely recovered from one type of human tumor in central Africa and in another type of human cancer in South China.

Viruses of this type may infect children at an early age, Berg suggested. It could be that they exercise their ability to remain silent until the stresses of modern living, air and water pollution, and unbalanced diets years later trigger activation. So why not vaccinate the children against virus infection before the viruses have a chance to initiate the long chain of events that may produce cancer?

Nobel Laureate Howard M. Temin, professor of oncology at the University of Wisconsin, also attended that Philadelphia birthday party. "Not on your life!" *Medical World News* quoted him as saying in reaction to Berg's proposal. Temin won a Nobel Prize in medicine in 1975 for the theoretical rationalization that led to discovery of reverse transcriptase, the enzyme that enables transmission of genetic information from RNA to DNA instead of over the more usual DNA-to-RNA route. Temin was working on replication of cancer-causing viruses and formulated his theory to account for events he observed in the laboratory. MIT's David Baltimore later demonstrated experimentally that Temin's theory was correct.

Vaccination of children against cancer might have any one of four outcomes, Temin said at the Fox Chase Cancer Center affair. Nothing at all might happen. If a viral cause of cancer were present in the child, the incidence of the cancer might be decreased. Or the incidence might be increased. Or "you may give the patient a completely different disease, as may have happened with last winter's flu vaccine."

Temin pointed out that results of manipulation of the immune system of the human body are difficult to predict. "It might take an entire lifetime to find out what the anticancer vaccine had wrought," he said; by that time "it would be too late."

Berg called Temin's objections to cancer vaccination illogical. "He certainly wouldn't stop polio vaccinations, and they were a calculated risk," said Berg.

It is already difficult enough to persuade people to accept vaccination even in those cases where safety and effectiveness are evident, Temin responded. "A shot against cancer seems so farfetched in its benefits, I wouldn't blame people for fearing the chimerical beasts of recombinant DNA."

Temin is a modest, soft-spoken man whose demeanor and competence give him wide credence among scientists. But Berg has a larger and higher speaking platform, with an official banner flying from it. Along with Stanley Cohen, he represents the United States on COGENE, the committee on genetic engineering of the International Council of Scientific Unions, and can be expected to use that opportunity to spread his ideas about the manipulation of viruses in relation to cancer.

The distressing implications of the Berg-Temin encounter were not discussed at the NIH's mid-December hearing on recombinant DNA research. In fact, the exchange between the two men was not even mentioned.

We have seen how the package of proposed changes in the NIH guidelines would affect the top as well as the bottom of the recombinant DNA experiment pyramid. But the story does not stop there. Another proposed revision would ease control over experiments in the middle of the research pyramid by removing the rigid physical and biological containment standards and substituting for them a sliding scale of values that would permit the lowering of physical containment in proportion to the increase in the biological safety provided by the organisms that are employed. All sorts of juggling by individual experimenters could be expected if this change were adopted. The consequences would be much greater than indicated on the surface. For Dr. Sidney Brenner, the British "disarm-the-bugs" pioneer who dominated the Asilomar conference in 1975, is reported ready to introduce a new bacterium that will rival Curtiss's χ1776. And Curtiss himself will soon report development of a second generation of the χ1776 family.

To cap the four very generous guideline relaxations just described, the NIH Recombinant DNA Molecule Program Advisory Committee recommended that the director of NIH be authorized to make exceptions to application of any guideline (including the one that absolutely bans experiments that might trigger cancer, spread resistance to antibiotics, or produce known poisons) whenever he feels that an exception would serve the public good. This would put tremendous power in the hands of a single individual. The strongest argument in its favor was that it was the only way to get started on the job that should have been started in 1974—actual experimenting to assess the risks of using certain types of DNA hybrids instead of taking the easy way out and guessing about it.

Whether considered individually or collectively, the proposed changes in the guidelines were difficult to understand. NIH admitted that in what was advertised as an instructive booklet (but was actually filled with vagaries). When politics rather than science was offered as a rationale for a recommended change or for the writing of a guideline in the first place back in 1975, some members of the Advisory Committee to the Director of NIH showed their annoyance. Politics was its job, the committee told the scientists on the NIH Recombinant DNA Molecule Program Advisory Committee, and it would be better for the scientists to stick to science rather than try to anticipate—usually in the wrong direction—how lay citizens arrive at their conclusions.

Although the Advisory Committee to the NIH Director encompasses within itself a wide range of expertise in other aspects of national and international life, most of its members were not sufficiently versed in science to challenge the findings of the NIH Recombinant DNA Molecule Program Advisory Committee on scientific grounds. Professor Robert Sinsheimer was an exception. He had ended his long tenure as head of the biology division at Caltech in order to assume the chancellorship of the University of California at Santa Cruz. But he was still keeping one hand in research in the viral genetic realm he had worked in all his adult life. And he came on strong in expressing his chagrin over the "extensive reliance on unpublished data" in documenting justification for the proposed guideline revisions. He declared himself "hesitant to endorse the numerous changes which correspond to a decrease in the level of containment." Georgetown University Law Professor Patricia King said she too was "terribly upset by the procedure for making the revisions." And another prominent Washington attorney, Peter B. Hutt, former counsel for the U.S. Food and Drug Administration, said the proposed revisions had been presented to the Advisory Committee with "undue, unnecessary and unseemly haste." Referring to the so-called Gorbach report of June 1977, Hutt said that "there was an obligation to make that data available to the public." The facts and figures should have been released, along with much other published information on which the guideline revision proposals depend, in ample time to allow the public to react.

The proposed new definition of recombinant DNA provoked pointed questioning of NIH Director Fredrickson. Who was to decide which species exchange DNA naturally and which do not? According to the text of that proposed revision, the NIH Director would make the choices with the advice of the NIH Recombinant DNA Molecule Program Advisory Committee. Everyone knew that the NIH director hasn't time to spend on such details. Fredrickson was asked how the job really would be done. A search of scientific literature would be undertaken, he said, and from it a basic list of applicable species would be drawn. Any recombinant DNA researcher who wished to add a species to the list could do so if he was able to present acceptable evidence to justify the addition.

Hutt was not satisfied with that explanation. What would constitute acceptable evidence? He couldn't find a definition in the guidelines. In what form was the evidence to be presented? The guidelines don't say. To whom would the evidence be addressed, how would it be tested, who would certify the tests and according to what criteria? What time schedules would be observed?

When Fredrickson confessed his unpreparedness to answer any of Hutt's questions, Hutt wanted to know why not. Hutt had asked similar questions regarding procedural matters a year and a half earlier, at the first meeting at which the Advisory Committee to the Director of HIH had addressed the governance of recombinant DNA. Hutt's plain implication was that if an honest attempt had been made to answer the questions when they were first asked, the ambiguities that later were offered as excuses for what happened at San Francisco would not have existed, and there would have been no justification for a violation.

To accent his annoyance over the neglect of the advice that Fredrickson had requested, Hutt did not merely refer to his original questions; he read them verbatim from a transcript of the proceedings of February 1976. The record of that time also contained a subsequent letter to Fredrickson in which Hutt had confirmed his view that the "present provisions in the guidelines do not adequately describe the intended requirements. The experience of the Food and Drug Administration . . . has been that general requirements are not adequate to assure appropriate activities. The more specific the requirement, the greater will be the adherence to it." In that same letter Hutt had reiterated: "It simply is not adequate to refer to acceptable practices, with the vaguest of generalized descriptions, since this is tantamount to no requirement whatever."

The term "guidelines" itself was misleading, Hutt wrote. They were not guides; they were regulations, carrying the force of law, and it would be folly to call them anything else simply to massage the egos of scientists to whom mere thought of regulation was a traumatic experience. If the realities of the situation were not faced forthrightly, Hutt predicted, the public sooner or later would become disillusioned and resentful and would then demand and get restrictions that really might suppress freedom of thought.

The movement to liberate recombinant DNA research from regulation reached a momentary crescendo on the second day of the deliberations of Fredrickson's select advisory group when Nobelist James Watson repeated his proposal of February 1975 that the NIH guidelines be scrapped. There was absolutely no need for them, he said. DNA recombinants were in no way dangerous. Government monies that could be going into valuable research were being frittered on Big Brother surveillance of creative minds. Watson said that he and the other ten members of the Berg committee had been "stupidly wrong" in calling for a limited research moratorium. The public hearing at which he was speaking was a "foolish" waste of time, energy, and cash. People who feared the consequences of unrestricted gene splicing had "no guts."

After Watson concluded his "apology to society," a high official of a university with which Watson was once affiliated remarked, "He was just being his usual extreme." No one present offered a contradictory opinion.

Fredrickson wound up the hearing without offering any defense to Hutt's "legitimate criticisms" except to say, "We are not regulating Campbell's soup."

It wasn't Campbell's. But he sure was making soup of his responsibilities, and it was not at all savory.

"I think it is a conflict of interest for the NIH to be both the sponsor, conductor, and regulator of this kind of research," he said. Earlier he had told the Stevenson subcommittee that the Center for Disease Control in Atlanta was best equipped to take over the task of enforcement of recombinant DNA research controls. "My own belief is that it would be to the maximum advantage of the country for a very simple package to be passed extending the existing guidelines to everyone."

No one within the range of his voice outwardly indicated appreciation of the ironic humor of the position he had carved out for himself. He did not want the public exposure and the sometime resentment that attaches to an openly proclaimed czar. But if the guidelines as amended were to be given legal sanction, which it was in his power to do, he would be a czar with all the conveniences of privacy.

Soon after the mid-December hearing, word went out that the NIH Recombinant DNA Molecule Program Advisory Committee would meet in February 1978 to hear a transcript of the testimony taken at the hearing and then decide whether to continue to support all the recommended guideline revisions, some of them, or none. Before February arrived, however, different word went out. The Program Advisory Committee meeting would be delayed until April.

29

WHEN THE 95th Congress of the United States reconvened for its second session late in January 1978, even the dullest observer in Washington could not mistake the signs that the recombinant DNA research controversy had carried popular faith in science to a tragic low. President Jimmy Carter's science adviser, Dr. Frank Press, was quietly exhorting his less sensitive scientific colleagues to stop demanding recognition as the Messiahs of the twentieth century, stop advertising the quick technological fix as a panacea for all of society's ills, stop proclaiming "breakthroughs" that had not yet occurred and might not even be imminent, and accept the reality that scientific and technological contributions to betterment of the human condition, though pregnant with historic potential, can be put to practical use only when they are emotionally, morally, and ethically acceptable to the public. Even as Press went about expressing these views, the air of the capital was becoming acrid with talk of multiple scandals in scientists' mismanagement of public monies.

Joseph Califano, the Carter appointee as Secretary of the Department of Health, Education and Welfare, was tightening controls on the $7 billion annual flow of grant and contract funds for which he is responsible. The General Accounting Office of the Congress was investigating expenditures by HEW and especially those of one HEW agency, the NIH, which the Stevenson committee of the Senate had exposed as lacking any desire to police regulation of the conditions under which $2 billion worth of experiments are done each year.

Most significant of all, Congressman L. H. Fountain of North Carolina, chairman of the House Committee on Government Operations Subcommittee on Intergovernmental Relations and Human Resources, which had thoroughly shaken up NIH administrative procedures in the 1960s, was preparing for hearings that would explore his concern "that the reforms we accomplished in the 1960s may not have endured." On the heels of the Stevenson hearings in November 1977, Fountain had told Deborah Shapley, a reporter on the news staff of *Science,* that the self-regulation supposedly practiced by universities, hospitals, and other research institutions "may be illusory."

Fountain subcommittee staffers said that the scope of his curiosity had widened to include not only NIH and its parent HEW but the Department of Agriculture and the National Science Foundation. According to information reported in *Science* by Shapley, the recombinant DNA research irregularities at the University of California Medical School in San Francisco and the Harvard University Medical School in Boston were widely believed to be symptomatic of sometimes much more serious abuses involving 50 to 75 percent of *all* grants and contracts paid from the public purse. Fountain was digging into the whole ball of wax.

Because *Science* is the journal of the American Association for the Advancement of Science, its reporters cannot afford to be strident. Often they cannot be entirely forthright. But Shapley came right out and said what she had to say: "No one is alleging—even in the most serious cases discovered so far—that scientists are using Federal grants and contract moneys, to buy mink coats, yachts, or private jets. But a number of important officials and groups are asking whether research funds are actually used [in] the way that the Congress and the Executive Branch think they are. . . . The activities range from out-and-out fraud to routine fudging of accounts, a practice that violates Federal rules but that seems nonetheless to be common."

Out-and-out fraud? Shapley recalled a notorious case that had been reported in the mass media as well as in *Science* in 1976. An NIH grant recipient, Leonard Hayflick, had grown human cell cultures at public expense. He was charged with selling them to other groups and institutions at a profit to himself of $67,000. NIH also alleged that he held sales contracts with a potential value of $1 million.

What kind of fudging of accounts had occurred? They were various. Apparently the most widely used, according to Shapley's sources, was a practice called "pooling." "Pooling" is the collecting, within a department, a school, or even a university, of the income from all grants and contracts into a common slush fund, which is drawn upon at random to meet all sorts of costs, leaving auditors helpless to determine which moneys were spent to accomplish the purposes they were intended for and which went toward objectives that might not have been approved had they been known in advance to the givers of the money.

"Pooling" at its worst was practiced in the case of Phin Cohen, who has filed a $1 million damage suit against Harvard, two of the counts being defamation of character and misappropriation of public moneys rightfully belonging to an untenured research captive. Congressman Fountain's staffers say that when Cohen first came to them, he confessed that while being trained as a scientist at Harvard he had been taught to believe that the subcommittee chairman was an archenemy of freedom of thought who specialized in removing the testicles of researchers who fell into his clutches. Cohen had been an assistant professor in the department of nutrition at the Harvard School of Public Health for many years but had never been given tenure. As a capable researcher, however, he was allotted laboratory space as long as he could bring in grant and contract support on the strength of his personal reputation.

Cohen charged that after his applications for grants were approved by NIH but before he had done any work, he was required to sign blank forms vouching for the way the money *had been spent.* Later, others in the department or the school would fill out the forms, reporting expenditures for items not in fact related to Cohen's needs. When he ran out of funds with which to continue the work he really was doing, he was told his grants were exhausted. Because he lacked tenure, his protests were futile until he risked the ruin his colleagues had warned him against and walked into the den of the devil from North Carolina, who nudged NIH.

NIH distaste for hunting down abusers of the public trust is so strong that the agency disguises its investigative unit of thirteen members by designating it as the Division of Management Survey and Review (DMSR). DMSR had no trouble in confirming the truth of Cohen's accusations; indeed, it stumbled onto irregularities in two other Harvard grants. Harvard offered no defense, repaying $132,349 to NIH so quickly that auditors for HEW decided to open an audit of all federal funds that Harvard receives—totaling about $400 million. That audit was still in progress when these words went to press in the spring of 1978.

Another interesting tidbit turned up in the first flurry of probing was an item of $87,500 spent for the breeding of experimental animals that were never used and apparently were destroyed. That money went to the Eppley Institute in Omaha, Nebraska, which has received a total of $18 million from the National Cancer Institute to test chemicals that may cause cancer. A rough dozen of Eppley's projects were found to have been approved on only a verbal say-so, reminiscent of the method used in the case of the misused recombinant DNA plasmid, pBR322, at San Francisco. The director of the Eppley Institute is Philippe Shubick, a member of the President's National Cancer Advisory Board, which has responsibility for overseeing the activities of the National Cancer Institute. False invoices and deposits of public funds in private bank accounts that drew thousands of dollars in interest were discovered in still another case.

Scientists complained that it was wrong to accuse them of cheating. They said government grants were or at least ought to be gifts, with very loose strings attached. Scientists were paid for thinking, not for accounting, and they resented having their brainchildren treated as missiles, safety pins, or diapers are treated.

On this point, Deborah Shapley quoted one government auditor of long experience:

> There has never really been a meeting of minds between the federal government and the academic community . . . professional staff who manage the funds as to what . . . they [the scientists] commit themselves to provide in return for research support. The academic community looks upon the effort to identify the . . . effort spent on each contract and grant as so much unwanted interference.*

The underlying cause for the violent attack on Senator Edward Kennedy's proposals to regulate recombinant DNA research was finally being dredged up. Kennedy had correctly identified the gene-splicing area as a good place to take hold of what he calls the "blank check policy" before the new technology had time to acquire too many slovenly habits. The hysterical rage that greeted his perceptions rose principally from the gene manipulators' realization that if responsible bookkeeping were to be required of *their* work there would be no excuse for the government's failing to apply the principle of accountability throughout the 75 to 90 percent of scientific and

* The absence of understanding was not nearly so total as this auditor indicated. In 1963 the American Institute of Biological Sciences was shut down completely for a day, its executive director was fired, and restitution of a quarter of a million dollars of public funds was demanded by the National Science Foundation following discovery of a long period of grant juggling.

technological effort that the American people pay for.

Whether by luck or through a stroke of silent statesmanship, the issue of *Science* that reached the journal's subscribers on the last day of the Stevenson subcommittee hearings in the Senate carried an essay on "The Code of the Scientist and Its Relationship to Ethics." The writer was Andrew Cournand, emeritus professor of medicine and special lecturer at the Columbia University College of Physicians and Surgeons and chairman of the editorial advisory board of *Man and Medicine: The Journal of Values and Ethics in Health Care.*

Adherence to the code had acquired unparalleled importance in the last quarter of the twentieth century, Cournand argued, because of the possibility of relating science to a "unified world-order of some type," "a relationship between the ethical stance of the scientist, *qua* scientist, and the problem of fostering humane socioeconomic development." But before the code could be so employed, its preservation had to be assured. It must be recognized that "science is now in a state of siege. Those who in the past have praised its contributions to human understanding and material well-being are now questioning many facets of the scientific enterprise. Some even go so far as to ask whether it does not contain the potential for destroying civilization."

Cournand recalled for his readers the four main points on which Robert Merton had constructed the code in 1942. First, the "quality of a scientific work should be judged on . . . scientific merits or significance alone." Second, "scientific works [must] be judged provisionally and only after the relevant evidence, so far as it can be brought together, is at hand." Third, "whatever the personal motives of scientists, the advancement of scientific knowledge must be the primary concern in the evaluation of scientific achievements." Fourth, an "individual scientist should share the knowledge acquired through his research with the scientific community, which has a right to that knowledge." In attempting to relate the code explicitly to the conduct of individual scientists, Cournand himself twenty-five years later, in collaboration with Harriet Zuckerman and Michael Meyer, had formulated the principal norms of science: honesty, objectivity, tolerance, doubt of certitude, and unselfish engagement.

In the essay in *Science* Cournand did not once mention recombinant DNA research. But the connection between his observations and current happenings in gene splicing were obvious. For example:

> In recent years the principles of integrity and of disinterestedness and selfless engagement in scientific activities have been occasionally infringed by referees and other readers of scientific reports. They have abused their privileges by selectively disseminating the contents prior to formal publication or by using the knowledge gained from the work for their own or their immediate colleagues' advantage. . . .

> An increased desire for publicity and personal recognition, without awaiting them as natural by-products of scientific achievement, is another of the dangers of the new situation in science. This is manifested, for example, by the now frequent practice of distributing the texts of scientific reports to representatives of the press prior to evaluation by scientific peers. It both bespeaks and threatens the loss of cohesion among scientists.

Perhaps, Cournand suggested, the scientific code was no longer as adequate as it once was. "Because of science's real successes, in increasing knowledge and understanding nature, and in becoming incorporated into technology in many spheres of life," he pointed out, "society has acquired more of a stake in what happens in science than it had previously." Foremost on his list of substantive modifications of the code that ought to be considered was:

> The code should explicitly take cognizance of the fact that the scientist is an individual who lives in a society which has ends other than the cognitive ends of scientists, and that the cognitive achievements of scientists do not always and necessarily serve these ends. Scientists themselves have multiple allegiances, both within the scientific community and outside it. They need norms to help them find the right balance among these allegiances. Indeed, scientists should follow their traditional code as far as possible in their transactions with the extrascientific realm of society. For example, they should show the same honesty in their dealings with nonscientists as in their dealings with scientists; they should show the same willingness to acknowledge the imperfections of their knowledge to laymen as they do to their fellow scientists. And in their criticism of scientists with whom they disagree on grounds which are a mixture of scientific and political considerations, they should show the same matter-of-fact disinterestedness and willingness to admit the good faith of those with whom they disagree as they ordinarily do within the scientific community.

Another area in which Cournand felt a need for the code to provide guidance concerned the "obligation of scientists to refrain from actions that would destroy science." As he summed it:

> This very general statement includes a continuum of possibilities, among which the perils of nuclear, biological and ecological warfare, and greatly diminished support by the public are not the least.

My own half century of experience in observing the impact of science and humanity on each other (to paraphrase slightly the title I gave to the now defunct monthly supplement of *Saturday Review,* which I created at the invitation of Norman Cousins and edited from 1956 to 1971) has led me, too, to hope for the ultimate emergence of science in the global leadership role that Cournand dreams of. And it seems to me that the realm of hybrid DNA molecule breeding would be the most natural site for initial implantation of the germ. As I see it, the crucial need of this dark moment in time is a father figure with courage to stand up and say that science has come to recognize its dependence on the society that shelters it and is ready to take its place as a responsible social force among the other social forces that drive civilization.

When I began the twenty months of investigation that produced this book, I saw one man who seemed to qualify for this Herculean responsibility. Before long I saw two, and finally three. All three appeared as witnesses before the Stevenson Senate subcommittee on the same day in November 1977.

First at the witness table was the president of the National Academy of Sciences, a biologist on leave from the faculty of Duke University, Dr. Philip Handler. It was he who underestimated the scope of the opportunity that came to him in the letter from the Gordon Research Conference on Nucleic Acids in 1973. It was he who approved the appointment of Stanford University Professor Paul Berg to chair an academy committee empowered to analyze the promise and problems of recombinant DNA research.

"Had we to do it over again," the academy head testified, "I would have made sure that Paul Berg's committee was appropriately weighted with some clinical scientists with long experience in epidemiology and in the handling of genuinely pathological organisms." That statement seemed straightforward enough. In the next sentence, however, Handler backed away from the implications of his admission of error, saying, "But I rather suspect that we would nevertheless be gathered here this morning and that the state-of-the-art [of gene splicing] would be much as it is today."

Berg was the next witness. He evidenced no intent to accede to the idea that he and the Berg committee that Handler okayed in 1974 had lacked either humility or sufficient concern for the democratic process. His thoughts ran in an opposite direction. "I am particularly concerned by the growing efforts and influence of the anti-science forces," he testified. "Deeply held and conflicting sociopolitical ideals challenge the traditional views of what science is for and how it should be done. Society desperately requires effective mechanisms for anticipating and evaluating the impact of scientific and technological breakthroughs. . . . In the recombinant DNA matter . . . scientists demonstrated that they could provide the early warning system for alerting society to the potential benefits of risks and their discoveries; accusations of self-interest, arrogance or even malevolence do little to encourage efforts of that kind. . . . Governments everywhere must seek better ways to encourage scientists' participation and the means to channel their input into the determination of policy."

A later witness in the Stevenson hearings that day was Professor Roy Curtiss III, in 1973 a scientific nobody but by 1976 established as the Luther Burbank of hybrid bacteria breeders. "As is well known," he told the Stevenson subcommittee in 1977, "there has been considerable criticism both inside and outside the scientific community that both the Asilomar and NIH guidelines were drafted by scientists to regulate a scientific technology that many of these same scientists were using or might later use. In hindsight, it is clear that these guidelines could have been drafted with much greater public input. . . . I think that the scientists who initially called for caution as well as those who contributed to the adoption of means for self-regulation acted in a responsible way . . . [but] we were somewhat naïve in not realizing that recombinant DNA activities would become such an important public issue. Continued criticism of science and increased distrust of scientists will have adverse consequences not only for science and scientists but also for society. It is therefore imperative that scientists be responsive to the concerns of society and entrust certain decision-making authority to representatives of society to ensure that the use of recombinant DNA technology will not have adverse consequences on the public health and welfare."

In my eyes Curtiss came out of the hearing room looking very much taller than anyone else in sight. In fact, he looked like a nominee for a Nobel Prize. He could get the prize only under the rubric of research in genetics. But it would really be the first Nobel Prize (the Peace Prize notwithstanding) ever awarded for one man's concern for the safety of his fellow humans.

Having declared my enthusiasm for Curtiss, I must document it. My readers are already familiar with him as a personality. Earlier in this book we saw how, while yet a nonentity, he had catapulted himself into the midst of the hybrid DNA molecule controversy in 1974 by writing an unusually long and unusually spirited letter to more than a thousand scientists around the world, including the members of the Berg committee. That remarkable communication suggested voluntary cessation of virtually all recombinant DNA research until "potential hazards can be assessed and means to cope with them established." That proposal did not meet with much acceptance, but it did assure him of an invitation to the International Conference on Recombinant DNA Molecules at Asilomar in February 1975. At Asilomar he tried to persuade his peers of the desirability of "disarming the bugs" and made an outright offer to undertake a "disarmament" project. It was a colossal gamble. Going on nothing more than a profound belief that biological containment of hybrid DNA molecules could be achieved through application of the old-fashioned virtues of patience, hard work, and moral purpose, he risked his professional future in the prime of his life, the reputation of his new laboratory at the University of Alabama Medical Center, and all the unspent grant money he had accumulated up to that moment.

Curtiss returned from Asilomar to his laboratory in Birmingham on March 1, 1975. He spent all of the following day calculating the most economical route to his objective. In the end, he laid out four alternative paths starting from four different strains of *E. coli* as breeding stock. On March 3, he called a meeting of his ten postgraduate associates (including his second wife, Josie Clark) and graduate students to tell them of the commitment he had made and to ask their help in fulfilling it as quickly as possible. He figured they could finish in a couple of months. But they worked incredibly long hours from that day until the twenty-third day of the following January, staying in the laboratory even on Christmas Eve, Christmas Day, New Year's Eve, and New Year's Day, before coming up with χ1776.

The number 1776 suggests that 1775 mutants were bred en route to χ1776. The actual number was 1,328, for Curtiss started with an earlier creation of his: χ536. From it the genealogy jumped to χ961, then to χ984, χ1276, χ1488, χ678, χ1697, χ1777, χ1820, χ1845, χ1846, χ1849, χ1855, χ1859, χ1864, χ1776! Curtiss had fixed χ1776 as his goal in advance (in celebration of the two-hundredth birthday of American democracy) and tried to trim his breeding program to fit it. But he underestimated the difficulties he would encounter and had to keep going up through strain χ1864 before falling back to the artificially vacated χ1776 niche.

Where had Curtiss miscalculated? From previous research, begun in 1974, he believed that most of the problem would be solved if he could breed a strain incapable of synthesizing its own diaminopemelic acid (DAP), a vital constituent of *E. coli's* outer coat. However, after he did breed a DAP-less strain, he found that its progeny mutated and recovered the coat-making ability. When the recovery gene was deleted, the new strain survived by making colanic acid, a sticky substance that held the bacterium's body together without a coat. The gene-controlling synthesis of colanic acid was finally knocked out, but when the mutant that lacked it was dying it could still conjugate with healthy *E. coli* and transmit infection to them.

The ultimate act of disarmament took away the bacterium's ability to make thymine, a component of its own DNA. Unless it were fed thymine in the laboratory, χ1776 simply could not live, much less propagate.

Although χ1776 was brought into being on January 23, 1976, Curtiss put the crippled creature through a long series of tests before reporting success in his mission to NIH Recombinant DNA Molecule Program Advisory Committee Chairman De Witt Stettin on March 30. More than a third of a million dollars had been spent on the job. The contract through which he recovered the money did not come through until July, and NIH Director Fredrickson did not certify χ1776 as safe for use in recombinant DNA experiments until December. But the significance of the Curtiss team's triumph was too great for news of it to be suppressed, even if anyone had wanted to suppress it, and invitations to describe his feat came to Curtiss in his lab in Birmingham from all directions.

Wherever he went, he was greeted with challenges. At a meeting of industrial microbiologists at Orlando, Florida, in February 1976, only a few weeks after χ1776 had emerged as a very weak and wobbly yet viable reality, he was asked what would happen to the DNA of his new creature during disintegration of the organism's body following death. In March, at public hearing on recombinant DNA conducted by the University of Michigan at Ann Arbor, someone wanted to know whether it would not make a difference if χ1776 should enter a human body through a cut in the skin of a hand, thereby making a direct entrance into the bloodstream, where the bacterium presumably would no longer need a cell wall to survive. At a scientific conference in the West German capital of Berlin in May, the question was: Does χ1776 die more quickly if exposed on the human skin? At the Tenth Miles International Symposium at MIT in June, doubts were expressed about the safety of χ1776's use in laboratories occupying the upper floors of buildings in which public medical clinics occupy the lower floors (a situation that applies to Curtiss's lab, among many others).

At each new appearance he made to introduce his overnight sensation to doubters, Curtiss was presented with at least one request for a fact he did not possess. He could not shrug off his unpreparedness because the questions being raised were the kinds of questions he had raised in his history-shaping letter of 1974. Progressively longer hours of his days were spent in the library, poring over scientific reports from the past. When they did not suffice, Curtiss telephoned specialists he had never consulted before. When they failed him, he set up experiments in his own laboratory. By the time September came, and the NIH Recombinant DNA Molecule Program Advisory Committee met and received the transcript of the closed August 31 meeting of NIH consultants, Curtiss had gathered data which he and other critics of recombinant DNA research had not previously been aware of, data which long before had established *E. coli* K12 as a very docile laboratory creature. In short, in designing the foolproof χ1776 Curtiss had come to the same conclusion the consultants reached: K12 itself was much less dangerous than the reputation the Berg committee had given it.

It is hardly ever pleasant for an individual to admit that he was wrong. It is especially painful if the error has swung its victim 180 degrees away from the truth as Curtiss had been swung in his 1974 letter to the Berg committee. Curtiss, however, is an exceptionally honest person, constitutionally incapable of evading or for long putting off open acknowledgment of his mistakes. Just how he would have gone about correcting the position he had declared in his globally circulated letter of 1974 if alternative options were open to him is an idle subject for speculation. For he was in fact caught up in the tidal wave of hysteria that engulfed the hybrid DNA molecule manipulators in April 1977. Senator Edward Kennedy was then proposing that gene-splicing research be overseen by a national commission, the membership of which would include one more layman than there would be scientists. Anyone who looked twice at this suggestion would surely see that if any knockdown, dragout dispute should develop on such a commission the scientists would have majority control by persuading only one layman that they were right.

Although Curtiss's parliamentary experience in the Oak Ridge City Council gave him a more realistic perspective on such matters than most scientists possess, he nevertheless joined the opposition to Kennedy in a letter addressed to Fredrickson on April 12.

Ostensibly responding to "Suggested Elements for Legislation," a report issued by the Federal Interagency Committee on Recombinant DNA Research, an entity President Ford had reluctantly authorized after sharp and repeated prodding by Senator Jacob Javits of New York, Curtiss said he was "partially [sic] reassured to find . . the suggested elements . . . more reasonable than some of the provisions contained in legislative bills previously introduced into the Senate and House." However, Curtiss added, he was "extremely concerned that . . . we are about to embark on excessive regulation of an important area of biomedical research" because of "fear, ignorance, and misinformation." He then presented, in nine single-spaced typed pages, a synopsis of interspersed data and extrapolated opinion leading to his "realization that the introduction of foreign DNA sequences into EKl and EK2 host-vectors offers no danger whatsoever to any human being with the exception . . . that an extremely careless worker might under unique situations cause harm to him- or herself." "The arrival at this conclusion," the Curtiss letter said, "has been somewhat painful and reluctant since it is contrary to my past 'feelings' about the biohazards of recombinant DNA research."

Curtiss did not then and does not now believe that he or his colleagues or the joys they share in science were or are in any danger of extinction because of a resurgence of Lysenkoism, Hitlerism, or even Kennedy-ism. But his letter to Fredrickson repeatedly referred to a vague fear of "excessive regulation," and used many of the anti-Kennedy lobbyists' favored clichés (some scientists would leave the country rather than accept dictation by the law; regulation would impose additional costs that would reduce funds available for research; legislation cannot prevent human error; licensure would unnecessarily hobble creative minds; inspections would undermine the authority of local biohazard committees); so the extremists had plenty of room in which to embrace him as one of themselves and use his letter in their propaganda. True to the twisted pattern of presentation that we have already observed, they did not take any note of his statement of the desirability of "additional data to substantiate [my] . . . assessment" or of his observation that what he said to Fredrickson was not the same as saying that an "individual with considerable skill, knowledge (most of which is currently lacking), and luck could not construct in multiple steps" a strain of *E. coli* K12 "that would be harmful to the human host."

We have seen how Kennedy's legislative proposals were stalled from April 1977 until the senator was bluffed into abandoning them in September. We have seen how events in San Francisco later demonstrated that Kennedy had been right in his main contention—that science had not yet learned how to govern itself in a way that was acceptable to the people who pay the bills or even to the permissive officials of NIH who issue the credit cards that accumulate the bills. We have seen how scientists summoned as witnesses before the Stevenson subcommittee in November supported measures which eight months before were being fought as oppressive. What emotional and intellectual turmoil did Curtiss pass through during the period before he appeared as one of those witnesses?

His April letter to Fredrickson put such heavy emphasis on his faith in the reliability of peer pressure and the dedication of local biohazard committee members that he must have felt great embarrassment when *Science* reported how miserably those trusted forces had failed in the laboratories atop Mount Sutro. But an evident wish to avoid involvement in that unhappy episode has moved him to withdraw from any extended discussion of his own motives. He will say only that "thorough reading and rereading of all the available information made clear to me what would be best for science and for the people at large." That "best" went into the statement he prepared at the request of Senator Stevenson.

"It would appear," the statement said, "that some have misinterpreted what I said" [in the February letter to Fredrickson]. "Although I am doubtful that recombinant DNA research or the commercial application of recombinant DNA technology will pose any threat to the public health and welfare or cause harm to other organisms in the biosphere, I certainly do not know this as a fact," Curtis explained. "Nor does anyone else have the substantial body of evidence to completely validate such a belief."

The assurances of safety given to Fredrickson in April, Curtiss declared, "only refer to recombinant DNA research with the *E. coli* K12 EK1 and EK2 host-vector systems. . . . I certainly didn't intend . . . my conclusions to be applied to recombinant DNA research with all conceivable host-vector systems nor to the use of recombinant DNA technology to design microorganisms for specific purposes. Furthermore, I was addressing myself to those experiments that are now permitted by the NIH guidelines for recombinant DNA research and would not want anyone to think that by the use of EK1 and EK2 host-vector systems it was necessarily safe to do those experiments that are now prohibited."

Because he had come to believe in the desirability of legislation to govern recombinant DNA research, he said, he felt obligated to offer some ideas that might be workable. There were ten of these:

One: "Relevant data and the methods by which they are obtained have to be fully evaluated before a consensus on the absence of risk . . . becomes valid."

Two: Until such evaluation is completed, the Congress should not assume "that all risks are absent, and . . . legislation . . . therefore unnecessary." On the contrary, it "would be prudent to adopt legislation to regulate recombinant DNA activities" whether publicly or privately financed "until such time as information is available to demonstrate that potential risks do not exist."

Three: The laws that the Congress does adopt should be so phrased as to "convey the importance of recombinant DNA activities and . . . indicate that the legislation is needed not because such activities are hazardous or ever likely to be hazardous but rather because sufficient information to justify that belief is not now available and . . . it is therefore best to be overcautious in providing for the public health and welfare and in protecting the environment."

Four: "Responsibility for formulation and implementation of regulations" should be vested not in the Director of NIH but in the Secretary of the U.S. Department of Health, Education and Welfare.

Five: Two committees should be established within HEW "to assist [the HEW Secretary] . . . in the formulation of these regulations."

One of these two committees should be similar to the present NIH Recombinant DNA Molecule Program Advisory Committee and should include scientists expert in the genetics and molecular biology of various host-vector systems; in human, animal, and plant genetics; in cell biology; public health; ecology; agricultural sciences; and industrial microbiology. The composition of the committee should be flexible so that scientists with additional expertise could be added as required for consideration of future applications of recombinant DNA technology. The committee should be responsible for the description of physical and biological containment categories, for specifying the levels of physical and biological containment for all uses of recombinant DNA technology, and for recommending approval or disapproval of new host-vector systems. This technical committee should also be responsible for confidential review of proposed activities that are inadequately dealt with in existing regulations and for formulating interim regulations to deal with those activities.

The other "assisting" committee assigned to the HEW Secretary should be responsible for approving or disapproving (but not rewriting) regulations drafted by the technical committee, for proposing means of implementing the regulations, for evaluating the impact of DNA technology on society, for making recommendations based on those evaluations, and "to make sure that appropriate experiments are performed to evaluate the potential risks associated with recombinant DNA activities." The membership of this committee should be "broadly representative of all segments of society," and "any members who are scientists should not presently be, or likely to become, engaged in recombinant DNA activities."

Six: The NIH guidelines in process of revision in 1978 should be specified as the interim regulations.

Seven: Federal law should preempt adoption of regulations by local and state governments "unless there was a proven need for local legislation."

Eight: "Institutions engaged in recombinant DNA activities should be licensed, and revocation of licenses should be the principal means to ensure compliance with the regulations."

Nine: The federal law should include sections dealing with the rights of employees and employers.

Ten: The law should include a "sunset clause" providing for termination of the legislation at a fixed time in the absence of reenactment.

It was abundantly clear from that decalogue that Curtiss's plea for legislative protection had a powerful positive orientation. A primary factor in his original involvement in recombinant DNA work had been his worry that antibiotic medicine was headed for ruin through thoughtless overuse and consequent spread of plasmids carrying resistance to therapy. Were such a calamity to materialize, what new form of medicine would arise to fill the void? Curtiss saw possible salvation in the genetics of the immunological system that is shared by all mammals and birds. And he now sees that dream moving toward reality.

The immunological system is, he told the Stevenson subcommittee, "one of the greatest mysteries in all of biology." The mystery resides in the mechanism that generates a multitude of diverse antibodies to fight disease. One antibody for each disease agent that attacks. "It has always been astounding that a given avian or mammalian individual could elicit the production of specific antibody molecules directed against between one and one hundred million different foreign proteins [that make up the invading organisms]," Curtiss said. "The amount of genetic information needed to accomplish this feat [if the orthodox belief that one gene controls one protein is accepted] would be in excess of the total amount of DNA present in the entire avian or mammalian genome [for direction of all the activities of life]. Although the complete story of this amazing . . . process is not yet available, it is clear that full understanding will soon be achieved." That prospect so excited Curtiss that he left his hearers to find their own ways of learning that recombinant DNA technology had made possible the discovery that a gene is capable of exercising not only one but many responsibilities and furthermore can operate at a distance and control the behavior of other genes, allowing pieces of DNA to be assembled like tool handles with optional heads. This seems to mean that a single gene can be a component of many different antibodies. "Five years ago," Curtiss exulted, "few if any biomedical researchers would have anticipated the solution to this problem in so short a time."

The unexplored genetic territory that gene splicers are heading into is far vaster than even most scientists suppose, Curtiss declared; using the host-vector terminology of the NIH guideline revisions, he outlined a rough verbal sketch of it for the enlightenment of the senators. Although χ1776, the standard strain of *E. coli* K12, and the two NIH-approved viruses, SV40 and polyoma, still offer great opportunities "to acquire knowledge [of the genome] not heretofore attainable," he said, "it is also evident that certain information might more easily *or only* [emphasis supplied] be obtained by the use of other microbial host-vector systems."

As the currently most promising candidates for the work that is to come, Curtiss named *Bacillus subtilis* 168 and *Pseudomonas putida* PpG1.

Curtiss identified *B. subtilis* 168 as a laboratory strain of a bacterium that has been used in genetics research for almost twenty years. *B. subtilis* is a nonpathogen, only very rarely associated with human disease and then only in patients already compromised by other diseases such as cancer. *B. subtilis* is an obligate aerobe, that is, it requires oxygen to live and would therefore be unable to grow in the intestinal tract of warm blooded animals because the environment of the animal gut is anaerobic, does not contain oxygen. *B. subtilis* lives in the soil and can form spores that are resistant to heat, freezing, and dessication. Spores are believed to survive in soil for hundreds of years. It was the peculiar ability to form spores that first attracted students of genetics and biochemistry to *B. subtilis* 168. Sporemaking is known to be controlled by genes. Those genes can be eliminated by mutation to assure the safety of *B. subtilis* for use in recombinant DNA experiments, diminishing the capacity to survive in the natural soil habitat.

"It is . . . likely that *B. subtilis* 168 host-vector systems . . . exhibit[ing] biological containment levels comparable to those" of *E. coli* K12 systems "will soon be available," Curtiss testified. Behind that statement stood the fact that a number of laboratories in this and other countries have been working on plasmids and bacteriophages that could make *B. subtilis* 168 a productive member of the hybrid DNA molecule family.

Pseudomonas putida PpG1, Curtiss's second nominee for the workhorse team in recombinant DNA research of the future, is a nonpathogen that cannot grow at the normal temperature of mammalian bodies. *P. putida,* like *B. subtilis* 168, is an obligate aerobe, a natural inhabitant of the soil micoorganisms as well as with other organisms that are recognized plant or animal pathogens. Most of the recombinant DNA work with *P. putida* has gone into the designing of suitable plasmids, some of which can maintain themselves in a stable state in many microbial species and therefore might pose a problem: transmission of hybrid DNA to other microscopic residents of the biosphere.

"Little work has been done, to my knowledge, on isolation of mutant derivates of *P. putida* PpG1 that would be restricted to laboratory-controlled environments," Curtiss said. "As with *B. subtilis* 168, there is little or no information about the ability of *P. putida* PpG1 or its derivatives either to survive in natural habitats or to transmit recombinant DNA to other organisms in natural habitats. Information on those points will certainly be needed to assess the relative safety of these new host-vector systems as they are developed."

Nevertheless, Curtiss expressed confidence that in time it "will be possible to provide levels of biological containment for *B. subtilis* 168 and *P. putida* PpG1 equivalent to those afforded" by the HV systems now certified by NIH and "thus allow the completely safe performance of recombinant DNA research with these new systems." "Still other host-vector systems" will be needed as the potential of DNA hybrids becomes more widely appreciated, he predicted. The quickest way to develop them, he suggested, is to start early "with strains of a given bacterial species . . . cultivated in a laboratory environment for a number of years; it might be expected that these . . . would be less able to survive in natural environments."

"In evaluating the safety of recombinant DNA activities," he said, "it will be useful to distinguish between research to acquire new [genetic] information as opposed to the use of that knowledge in connection with recombinant DNA technology for the production of a specific product or for the development of a specific process that will be applied in solving some problem of society." He held forth none of the glowing promises that the Berg committee had persisted in making to justify its constant pushing for speedup of recombinant DNA experimentation. He did not even mention bacterial mass production of human insulin, genetic treatment or cure of cancer, adaptation of food crops to allow them to draw the nitrogen they need for growth from the air and so end their dependence on fertilizers, manipulation of the photosynthetic process by which green life pulls energy directly from sunlight, or quick cleanup of oil spills that pollute the seas and bordering beaches. Instead, he specified that he was speaking "in a futuristic sense" when he turned from his own preoccupation with the human genome to consideration of possible applications of recombinant DNA technology to the immediate needs of society. "Ultimately," he said, he believed that gene splicing "will . . . help society increase availability of food and energy, improve health and cope with pollution." But his unspoken implication was that much time, sweat, and tears would be spent on the road and that the effort would be less traumatic compared with the preparation going on now and in the years just ahead. Again, it is "conceivable that other microbial host-vector systems might be more utilitarian than *E. coli* K12 host-vectors" in view of the enormous variety of species of microorganisms that mediate the processes maintaining the mobility of modern civilization.

Curtiss's prepared testimony before the Stevenson subcommittee had been written, typed, and Xeroxed long before the hearing began. He knew from the agenda that the subcommittee staff had distributed that Handler and Berg would testify before he did. But he did not suspect that they, in their eagerness to shore up the extremely weak record of progress toward the practical applications of recombinant DNA technology that had been promised years before, would take the extraordinary step of announcing the results of a not yet published experiment. In that experiment Herbert Boyer and others had persuaded χ1776 to reproduce somatostatin, a quite small human hormone that plays a part in the regulation of growth. In total innocence of the fact that he would be raising the curtain on a scientific display of counterpoint, Curtiss had chosen the manufacture of a human hormone (any hormone) to illustrate the difficulty and time-consuming nature of the process of putting recombinant DNA technology to practical use "as economically as possible."

It would of course be expected that the task would begin with an organism containing the appropriate human hormone gene, inserted in a plasmid. Once the gene was isolated and purified, it could be hooked into the plasmid with the help of a restriction enzyme that recognized a convenient sequence of the nitrogenous bases representing the four letters of the genetic alphabet—A, C, G, T. The plasmid could then be introduced into χ1776 after either the plasmid or χ1776 had been modified to make four further steps possible.

The first step would be to bring about expression of the controlling gene so that the hormone would be faithfully synthesized by χ1776.

The second step would be to prevent χ1776 from degrading the hormone after χ1776 manufactured it. (The *E. coli* K12 family of bacteria, to which χ1776 belongs, habitually degrades proteins that are not useful to the bacteria. So far as is known, hormones are a type of protein of no help to bacteria.)

The third step would be to teach χ1776 to secrete the hormone into the culture medium in which the χ1776 bacterium is growing, thereby facilitating recovery of the hormone free of bacterial contamination.

The fourth and last step would be to adapt χ1776 to live in a fermentation vat filled with an inexpensive culture medium "capable of yielding maximum amounts of hormone and a minimum mass of bacteria."

After hearing the unexpected testimony of Handler and Berg—which, although fuzzy on details, indicated that Boyer and his associates had accomplished only the first two of the four steps that Curtiss said would be essential to a truly economical operation (one that would have a reasonable chance of reducing the cost of the hormone to the person who needs it)—Curtiss had full opportunity to amend the statement that he had submitted earlier to the subcommittee in multiple copies. But he let the testimony stand in its original form.

I telephoned Michela Reichman, the University of California news director on the San Francisco Medical School campus, to learn more about the Boyer team's "astonishing" feat, as Berg had called it. It didn't seem at all astonishing to me because it was exactly what Boyer had previously told me he would be able to do. But I did want to know how he did it. Michela is a very dependable communicator, who deserves her reputation for truthfulness. She said she couldn't discuss the experiment because the report on it hadn't yet been published, and she didn't know when or where it would be published. She could confirm only that Handler and Berg, insofar as they had gone, had been correct in their remarks. She promised to send me a copy of any description that was issued as soon as it became available.

Eight days after Handler, Berg, and Curtiss had testified, Boyer appeared before the Stevenson subcommittee along with Rutter to explain the violation of the NIH guidelines in the case of the plasmid pBR322. Michela was with them. She still couldn't talk about the somatostatin experiment, but the fact that she was there, plus the fact that she didn't ask to be excused from future discussions on the ground that the experiment didn't fall within her bailiwick, left me confident that the university laboratory had been involved in the work. I was therefore surprised to hear Boyer, in the midst of his testimony, volunteer the information that the somatostatin experiment had been wholly financed by Genentech, a small privately owned company that expected to profit from the project. Boyer was one of the organizers of Genentech, and one of its major owners. Senator Stevenson seemed as taken aback as I was. Throughout the remainder of that day and the following day, he questioned many witnesses about the patent rights that prevail under different kinds of circumstances.

When the report of the Boyer team was published in *Science* several weeks later, I got the promised press release from Michela. Written on the standard University of California-San Francisco press release form, it announced a press conference to be held December 1, 1977, in the Biltmore Hotel in downtown Los Angeles under the auspices of the City of Hope National Medical Center. The release quoted Handler's premature advertising plug as a come-on for newsmen, and told me enough to give me an understanding of how the experiment had been done.

Four scientists on the City of Hope staff—Dr. Keiichi Itakura, Dr. Arthur D. Riggs, Tadaaki Hirose, and Roberto Crea—had collaborated with Boyer and two of his university laboratory associates—Herbert Heyneker and Francisco Bolivar (one of the designers of the pBR322 plasmid) in the somatostatin experiment. Itakura had invented a new method for making artificial genes and had used the method to construct a synthetic copy of the gene that controls the production of somatostatin in the human pituitary gland for distribution through the brain and gut. This artificial gene contained only eighteen amino acids. Genetic sequences (nitrogenous bases of DNA representing the genetic alphabet letters A, C, G, and T) were then added to the gene so that χ1776 would replicate the resulting bundle of hereditary information. Next, the bundle, which χ1776 would not have recognized as being intended for replication, was inserted into the middle of a natural *E. coli* K12 gene that χ1776 would recognize. The artificially impregnated natural gene was then inserted into a pBR322 plasmid with the help of a restriction enzyme that knew the right combination of As and Ts, Cs and Gs to make readable three-letter genetic words. pBR322 in turn was inserted into χ1776, and χ1776 replicated the genetic bundle that had the somatostatin inside.

At that point, a modified form of Curtiss's step one had been achieved. There was no chance that χ1776 would degrade the hidden hormone because it didn't know the hormone was there. So a modified form of Curtiss's step two had been achieved also. However, instead of going on to execute Curtiss's step three by teasing χ1776 to secrete the somatostatin into the culture medium in which χ1776 was growing, Boyer's team killed χ1776 and took the entire genetic bundle from it. According to Boyer, the bundle is inactive in the human gut and is not known to have any ill effect on humans. So there was no apparent biological hazard in the handling of it. Finally, the somatostatin was snipped out of the bundle with cyanogen bromide. All the work was done in P3 laboratories as required by the NIH guidelines. Although no one said as much, it was clear that at least some of the work took place in the same P3 laboratory where the pBR322 plasmid scandal broke.

At the City of Hope press conference, Boyer told newsmen that the amount of somatostatin that had been produced by χ1776 represented 0.03 percent of χ1776's weight. He called that a "significant"

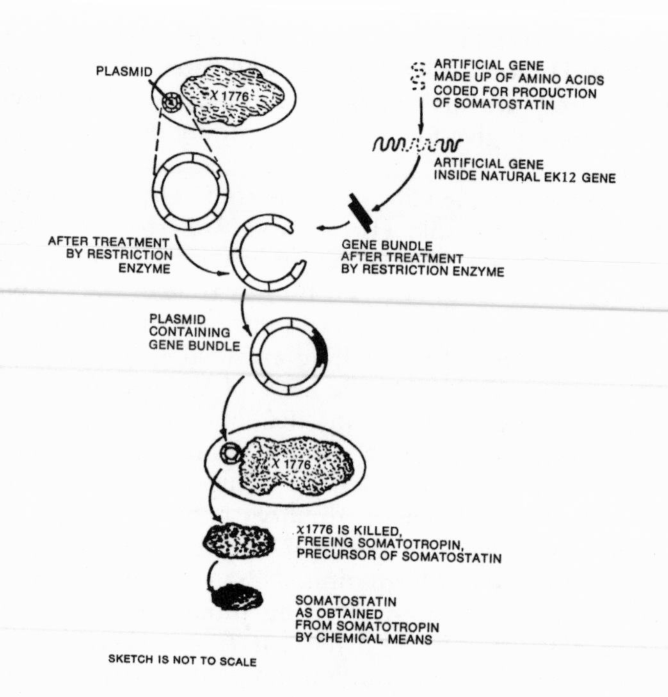

fraction. It was one ten-thousandth of the whole bacterium. "The man on the street can now finally get a return on his investment in science," Boyer said, adding that techniques "virtually identical" to those used in the experiment "could be used safely . . . to produce complex biological substances varying from insulin and other hormones to the enzymes used in industrial processes."

Boyer's optimistic forecast apparently was based on the earlier Boyer-Rutter-Goodman feat of replicating the gene that controls production of the insulin hormone in the pancreas of the rat. But only the gene had been reproduced in that instance. χ1776 did not express the gene, that is, the rat insulin itself was not manufactured.

Whatever the basis for Boyer's remarks, they were disagreed with even before he uttered them. Immediately following the unorthodox Handler and Berg sales pitches, *Nature,* in an unusual "backgrounder" report, had cast doubt on the practical meaning of the somatostatin exploit. As for somatostatin itself, the "backgrounder's" anonymous author noted that somatostatin affects the action of other human hormones and inhibits the aggregation of blood platelets, which are important to blood clotting and the prevention of hemorrhage. Even if these drawbacks to widespread use of somatostatin in medical research were to be overcome, it would remain true that the making of synthetic genes is practical only if the genes are small. And that would rule out insulin because in comparison to somatostatin, the insulin hormone is huge.*

On the morning after the City of Hope press conference, the *San Francisco Chronicle* carried a commentary signed by science correspondent Charles Petit, who had dug into Genentech's genealogy. Robert Swanson, the thirty-year-old president of the little company, had been an investment analyst for Citibank of New York. In January 1976, he telephoned Boyer and proposed formation of the company, in which Boyer is now a vice-president. The chairman of the board is Thomas J. Perkins, of the San Francisco investment firm, Kleiner and Perkins. Swanson says there are "about six" principal investors. Roughly a quarter of a million dollars went into the trial run on somatostatin.

In return for that money, Genentech expects to receive from the University of California-San Francisco and the City of Hope an exclusive license to make hormones under patents that have been applied for on the process and the organisms involved. Genentech also hopes to swing a similar deal on an insulin production method that Boyer says will be "tens or hundreds of times" lower in cost than are chemical production systems or methods that depend on isolation of the hormone from animal pancreases, and although spokesmen for the Pharmaceutical Manufacturers Association and a small California company named Cetus (Joshua Lederberg is a consultant to it) had told the Stevenson Senate subcommittee that the marketing of human insulin probably could not occur in less than five years, Genentech president Swanson said Genentech might turn the trick within a year. The firm was dickering for a property in south San Francisco on which to build its own laboratory and possibly a factory.

Petit said no one he interviewed was willing to buy the Boyer-Swanson prospectus. They all called Genentech a "risky venture," giving as their reasons the uncertainty about legal controls and "confusion" over how patent laws will be applied to new life forms created in the laboratory. So far as I can discover, there is no confusion on that particular point. Hybrid seed corn is a life form that was created by laboratory methods, and it was patented many years ago by the Research Corporation, a nonprofit organization with long service as owner, licensor, and administrator of the rights to

* Other qualified scientists were not satisfied that the somatostatin that had been produced was a human hormone. A number of mammals secrete somatostatin, but whether their forms of the hormone are the same as the human form is uncertain.

inventions of basic importance. More recently, the Upjohn Company has received a patent on a biologically pure culture of a microorganism. Recombinant DNA technology was not used in that development, but will be applicable to it.

Putting aside this one point and looking at the patent system as a whole, the system is a shambles. Congressman Ray Thornton, of Arkansas, chairman of the Science, Research and Technology Subcommittee of the House Committee on Science and Technology, has offered a bill, jointly with Texas Congressman Olin Teague, to establish a uniform federal system that will provide sensible utilization of the results of federally sponsored scientific and technological research and development.

Genentech's situation did not fit cleanly into the overall patent discussion. Boyer's own argument was identical to the one Thornton espoused. But Petit found that "in using publicly owned facilities at UC for some of the work" Genentech had "raised eyebrows among some academic researchers in the recombinant DNA field. Boyer's personal investment in the profit-making firm has caused a few colleagues to wonder whether he can continue to take part in the debate over public policy and regulation of genetic engineering and related fields." At Stanford, where Stanley Cohen had enlisted Boyer's participation in the first plasmid-splicing experiment in 1972, Petit found some researchers a "little disappointed that Herb is trying to commercialize his work so quickly." Although Berg had taken an enthusiastic part in the unorthodox advance advertising of the somatostatin experiment, Berg said commercial involvement "is just not to my taste. This isn't to criticize Herb particularly, but I just can't see it."

Curtiss was in San Francisco at the time for a San Francisco State University symposium on ethical and policy problems surrounding recombinant DNA technology. He said to Petit: "If you are directly involved with patents and proprietary information, and you work full time for a university, you're constrained in what you can say about your work." Also, Curtiss said, when anyone in such a situation conceived an idea, he might be tempted to ask himself: "Is it something I could pass on to our little company?"

Curtiss's reaction was especially interesting because he has applied for a patent on χ1776. He removed himself from direct involvement, however, by making the application through the nonprofit Research Corporation.

A Stanford patent attorney, Neils Reimer, dismissed the patent issue as a tempest in a teapot. It was he who took the initiative in applying for patents on the original work of Cohen and Boyer. "The government wants patents on university work," he said, echoing

Thornton's view, "that's how things find their way to application." Stanford has had some second thoughts, however, and recently asked Reimer not to process the Cohen-Boyer application further until public policy issues surrounding recombinant technology are settled.*

The longer Congress refrained from acting, the more chaotic the situation surrounding recombinant DNA research became. Everyone involved was jumpy, and in the general nervousness very few were taking accurate readings of the pertinent signals of what lay ahead. Almost no notice was paid, for example, to the concluding paragraph of a self-congratulatory guest editorial that ASM President Halvorson wrote for *Science* late in October 1977 after Senator Kennedy had been driven from the gene-splicing field. Here is that paragraph:

> There is need now for a new era of openness in the dialogue between the scientific segment of our nation and those who represent us in the Legislative Branch. No longer dare we [scientists] flaunt our perceived power or underestimate the genuine efforts of concerned citizens to protect themselves from risk. The treasured freedom of scientific inquiry can be rapidly eroded if we "come on too strong" with self-serving pronouncements and over-zealous protective positions. It is time [for scientists] to speak, but it is also a time [for them] to listen—carefully.

The jitters epidemic spread to such proportions that just two weeks before the 95th Congress was due to return to Washington for its second session, Dr. Philip Abelson, the editor of *Science* and a distinguished scientific investigator in his own right, published one of the strangest editorials that I have ever seen in that journal. It was a fervent, almost an abject, plea to researchers of every stripe to behave as though they were in church until Congress finished its business and got out of town again.

Citing Stanley Cohen's report of his latest experiment as though its controversial content had been confirmed in the customary way, and without noting the devious tactics that had been employed by the science lobby of 1977, Abelson credited the lobbyists with holding off a "threat to the freedom of inquiry." He accused "some of the pro-

* In a letter dated March 2, 1978, NIH Director Fredrickson confirmed Stanford's exclusive right to a patent on the Cohen-Boyer technique and the concomitant privilege of licensing use of the technique to persons or corporations who abide by the NIH gene-splicing guidelines. NIH thus provided a means of bringing privately funded gene splicing under the guidelines. However, Fredrickson said, "use of the institutional patent agreement as a means of obtaining compliance with the NIH guidelines is not an adequate substitute for legislation." A Stanford spokesman said the university had made no decisions on licensing, explaining, "We don't want to do anything that would interfer with legislation relating to recombinant DNA research now before the Congress."

ponents" of "restrictive legislation" of bad faith in their professions of fear of "threats of recombinant DNA." Their declared target was only "ostensible," he said; actually, they had "broader objectives." The implication was that those objectives were not laudable, but he did not identify them or the people whose efforts he was deriding.

The escape that the lobbists had managed to contrive might, however, Abelson warned, "prove to be only temporary," for "some kind of legislation seems likely." "Irresponsible acts by individual scientists could be very damaging" in the delicate waiting period, he said; "Almost anything can happen depending on the public mood of the moment." When the legislation finally is "adjusted in a conference meeting of the House and Senate, strange provisions can enter that bear little relation to the original bills. A major hazard is that during the crucial moments . . . , news will come out of some irresponsible act by a scientist engaged in recombinant DNA research. This need not be an act of substance. . . . Failure to complete some paper work could draw censure. . . . Recombinant DNA research . . . is highly competitive. Workers are under temptation to take short cuts. But they should behave as if their every act is under scrutiny, for indeed it is—by assistants, colleagues, or competitors. A scientist who furnished the pretext for restrictive legislation could count on the ill will of many of those he or she wants most to impress."

I read that editorial several times, in total disbelief. I had known Abelson's work for many years, had respected him greatly, and as a magazine editor had published some of the less popular positions he had had the courage to take over the years. I wondered why he was hiding behind innuendos about the motives of unidentified advocates of legislation. It would have been far more in keeping with his character to name names and specify the objectives he was attacking.

In any case, Abelson was not in a strong position to invite retaliation. Under his editorship, *Science* had killed a news report on the National Academy of Sciences forum on recombinant DNA technology, where the gene-splicing establishment suffered its worst repudiation. None of the damning transcript of the Stevenson subcommittee hearings that is reproduced in this book had been published in *Science* at the time of the hearings, although *Science* reporters attended the hearings. The Gorbach report had been mentioned favorably in an editorial signed by Abelson, with no commentary in the news section of *Science* to tell the circumstances of the report's dissemination. The same editorial had uncritically accepted Cohen's unreported "new work" without giving *Science* subscribers (of whom I am one) the benefit of a factual report on Cohen's political maneuverings and their consequences. These were not picayune failings, interesting only to scientists. The pages of *Science* are a primary source of usually dependable information for science writers on the staffs of newspapers, magazines, radio, and TV stations. If the story is twisted or truncated in *Science* the understanding of the outside world will be twisted or truncated, too.

I did not see how Abelson could escape a public dressing down. The numerous fellow scientists and environmentalists whom he had offended were the least of his worries. He had also jarred the sensibilities of some senators and congressmen. To me, the only question was which of them would open fire first and how soon.

By a curious twist of justice, the counterattack came in a two-day symposium on recombinant DNA technology at the mid-February 1978 annual convention in Washington, D.C., of the American Association for the Advancement of Science— *Science*'s publisher. A scheduled speaker at the symposium was Congressman Richard L. Ottinger of New York, a member of the Rogers Health and Environment Subcommittee of the House Committee on Interstate and Foreign Commerce.

Edward Kennedy's prominence in the Senate had overshadowed Ottinger's early involvement in the gene-splicing row in the House, where he introduced a resolution in January 1977 calling upon NIH to broaden its guidelines to cover privately financed recombinant DNA research by the following April 30. In February 1977, after the resolution had been bottled up, Ottinger introduced a bill to instruct the Secretary of HEW to issue guidelines applicable to privately financed and publicly funded research within ninety days of the bill's enactment into law. No patents would be issuable if the guidelines were transgressed. Researchers would be "strictly liable, without regard to fault, for all injury to persons or property" caused by their experiments. All gene splicing would be licensed, with the facilities subject to inspection and their operators open to a one-year jail term and $10,000 fine for each day of violation of the regulations. Laboratory employees who reported infractions of regulations would be protected against dismissal or harassment.

It was not until March 1977 that Subcommittee Chairman Rogers and others on the subcommittee submitted a milder bill that cut infraction penalties in half and allowed local biohazard committees to issue licenses monitored by HEW. Ottinger did not support that bill until it was revised in June 1977 to include provision for appointment of the special commission to study the pros and cons of recombinant DNA technology for two years and then report recommendations to the President, Congress, and the people.

Congressman Harley O. Staggers, of West Virginia, chairman of the full Commerce Committee of the House, used Cohen's "new work" to prevent Rogers from getting the revised bill onto the House floor during the first session of the 95th Congress. In the ensuing eight months prior to the convening of the February 1978 AAAS symposium, Burke Zimmerman, a former Environmental Defense Fund scientist who had moved into a key place on the Rogers subcommittee staff, tried again and again to find language that would be weak enough to please Staggers and yet say something meaningful about the gene-splicers' responsibilities to the American people.

Zimmerman took the latest draft of the Rogers bill with him to the AAAS symposium. It was already common gossip in the scientific community that lobbyists for Harvard University had insisted that in order to obtain their approval the proposed legislation would have to free Harvard from the vigilance of the city of Cambridge. So no one was surprised to hear Zimmerman say that the bill now had five principal provisions: (1) extension of the NIH guidelines to cover privately financed as well as publicly funded gene splicing for a period of two years; (2) authorization for the Secretary of HEW to make exceptions to the guidelines for whatever causes he and an advisory committee considered to be for the public good; (3) exemption from the NEPA requirement for environmental impact statements on guideline revisions; (4) preemption by the federal government of state and local community rights to intervene; and (5) an interim two-year study of recombinant DNA technology's benefits and risks by the special commission Ottinger had demanded as his price for supporting the legislation.

Zimmerman said that the text of the bill had been circulated among gene splicers around the country and had their approval. He also said that all "appropriate" members of Congress had been consulted and had agreed to the text.

Zimmerman's last remark was a strategic mistake. He had not consulted Ottinger on the preemption provision. With a great deal of justification, Ottinger considered himself very "appropriate." And when his turn came to speak, the New Yorker repeated a statement he had introduced into the *Congressional Record* the day before, in which he declared his opposition to the preemption clause. He said he felt sorry for Harvard if its ability to compete in the gene-splicing competition with Chicago and California depended on Harvard's freedom to cut corners across the beefed-up guidelines that the City Council of Cambridge had so carefully worked out. Regardless of the university's reason for demanding preemption, the congressman said, Harvard was taking the wrong approach to its problem. Instead of trying to pressure the federal government into pulling the rug out from under the city council, he said, Harvard should present the council with a persuasive case for the school's point of view and, if that failed, should put Harvard representatives before the voters of

Cambridge as candidates for seats on the council and work from the inside in a democratic way to change the law.

Ottinger said he had been given "a fascinating insight into the arrogance and anti-democratic spirit of which the 'established' scientific community . . . [is] capable. Those who have lobbied so hard to prevent the Congress from taking steps many of us are convinced need to be taken . . . apparently think themselves omniscient and infallible. 'We're the experts,' the saying goes, 'and you can't possibly understand whereof you speak.' I resent that . . . extremely, and the American public will destroy you if you maintain that attitude."

Abelson's *Science* editorial of a month earlier could not be read apart from "the frequent villification . . . heaped on the heads" of gene-splicing critics "by scientific establishment members," the angered congressman continued. "The warning that any scientist who, during the public debate, makes a mistake which heightens public interest in the DNA issue and hence lends support to more proscriptive legislation will be unpopular with his colleagues should be viewed together with other suggestions that have been made that scientists who themselves urge constraints may similarly be alienated from the scientific mainstream." Ottinger saw a parallel between these events and "the horrendous treatment by the atomic establishment—in and out of government—to make 'non-persons' out of such concerned scientists as Irwin Bross, Arthur Tamplin, John Goffman, Robert Pollard, Philip Mancuso and Ernest Sternglass, to name a few. . . . I sincerely hope that this time researchers into the problems and hazards of DNA will not find their research stifled."

In defending his own actions, the New Yorker declared it was "not only a right but a duty" for him to "err on the side of conservatism in protecting the public from danger." His one-word description of the recombinant DNA technology control bill, as it stood: "miserable."

Pamela Lippe, spokesperson for Friends of the Earth, heard Ottinger's words with satisfaction, and joined him in criticism of the bill. She charged that most of the safeguards that environmentalists had tried throughout 1977 to incorporate into the legislation had been omitted.*

A second recombinant DNA research symposium at the AAAS

* Although Ottinger and others within the Commerce Committee opposed the preemption clause, it was approved by a vote of 17 to 6 just before Palm Sunday 1978. By the usual rules of the House, the measure went next to the Thornton subcommittee of the Committee on Science and Technology, which held two days of hearings in April. A Senate bill was due to be marked up on May 4.

meeting was enlivened by a question from the convention floor to Roy Curtiss. What did he think of Stanley Cohen's manipulation of an unpublished scientific paper for political purposes? "It was one of the most imperious, despicable pieces of political science that I know of," Curtiss replied. He agreed with others on the symposium panel that Cohen had not observed K12 behavior under natural conditions, but even if it should turn out that that behavior did occur naturally it would provide an answer to the wrong question. The right question is how do we protect ourselves against possible consequences of stepping up mutation rates in the laboratory far above the slow pace with which nature proceeds? That consensus was a tribute to Sinsheimer, who had been saying all along that he was not philosophizing but talking down-to-earth common sense. Frank Young, of the University of Rochester, spoke wittily of his hope for a return to old-fashioned humility in science. He said that as far as he was concerned any scientist who thought he was doing something that nature didn't know how to do was like Little Jack Horner in the nursery rhyme, who thought himself bright because he could pull a plum out of a plum pie.

30

I AM GOING to talk about myself a little bit because I think it is necessary to know, if you can, what kind of person it is who comes to the conclusions and says the things that I propose to say. I was and remain full of what might be called religious wonder and awe at the universe and all that is in it, including human beings and all living organisms. And I had and still have an abiding, if naïve, faith that men and women can bit by bit go on indefinitely decreasing the immensity of our ignorance by scientific investigation and by humanistic insights.

It has not been easy to maintain that faith, or to recover it after lapses, in view of the bad times that Chargaff and Wald and I and others of our generation have lived through—the Great Depression, the Hitler years, the Lysenko period, the McCarthy period, the A-bomb, Vietnam, Watergate, and now anti-intellectualism. Some experiences of the last few days haven't made it any easier.

I share with George Wald and many others the conviction that scientists can convey to nonscientists, and to those who consider science incomprehensible, the major scientific facts and principles and the spirit and faith of science. I accepted my assignment here because I thought I would be going through a comparable exercise.

Those words are borrowed from Indiana University's Distinguished Emeritus Zoology Professor Tracy M. Sonneborn, who has been teaching genetics for forty-five years and who was, as he puts it, "sensitized to the societal implications" of his teaching fifteen years ago to the extent of staging the first symposium ever designed to anticipate the impact of genetic engineering. The above-quoted words were addressed to the National Academy of Sciences forum on recombinant DNA research in March 1977, and I repeat them here because they say so well what I need to say to identify myself.

I belong to Sonneborn's generation. My biases occasionally differ from his because he is a scientist and I am a lay observer of science with a single academic credential, an honorary doctorate of science, to match against his impressive string of well-earned symbols of learning. But I disagree with him on only one item in his catalog of the unpleasantries experienced by our generation, that is his characterization of the present as a time of anti-intellectualism.

Sonneborn had been asked to synopsize the first two days of the two-and-a-half-day forum. His remarks are equally applicable to the many months of work on this book. So I shall again borrow from his singularly graceful yet down-to-earth prose, as Mrs. Betsy Turvene recorded it in her expertly edited volume of the forum proceedings:

> The public is vitally concerned with some areas of scientific research . . . [and] should be appropriately informed . . . in time to fulfill effectively its responsibility in arriving, in discourse with the scientists, at judgments and actions about choices and regulations. . . . No matter what one's prejudices or biases may be about this or that interest group . . . , all those that think they have an interest have to be given full opportunity for input. . . . If we take democracy seriously, base has to be touched with all. . . .
>
> One thing depressed me more than any other. . . . Scientists, by their training and experience, [are expected to] . . . have respect for facts . . . , to grasp probabilities, and . . . [to be]willing and able to listen to . . . others who have a different expertise, and even occasionally to change their minds . . . [But here] scientists came with their minds made up, not yielding on anything of importance. I believe they all have their conception of the public good passionately at heart, but sometimes they seemed . . . to have their . . . ears tightly closed, and to lean on sometimes demonstrably false facts, or to remain deaf to cogent reasons.
>
> It is . . . unworthy of scientists to impugn motives or run down opponents personally. The purely scientific issues can be scientifically settled only by facts; or, when facts are lacking, by getting them if possible; and if not possible, by the rule of reason. Until scientists discussing the present scientific issues do that, they can only confuse . . . and frighten the public, and feed the burning fires of anti-intellectualism.

> It is my impression and my faith that nonscientists can at least sometimes see through the obliquity of us scientists and come to a workable consensus, as it seems to me the lay committee at Cambridge did.

Nonscientists deserve Sonneborn's declaration of faith. Their ability to see through the murk that scientists too often generate around themselves and their doings is present not only sometimes but much of the time. The process of penetration can be made tedious, however, by dust that scientists, sometimes intentionally and sometimes innocently, throw into the air. Sonneborn innocently threw some at the academy forum in his several references to anti-intellectualism. A few small bonfires probably are being kept alight here and there by anti-intellectuals, but a conflagration? There is a spreading flame of desire among the citizens of the United States and of other nations as well to share in the intellectual ferment. The well-educated fraction of the earth's population is vastly greater than it has ever been before, and it has acquired too much sophistication to accept the half-truths, the authoritarian posturings, and the mumbo-jumbo that characterized too much of science in the past.

I was startled when Sonneborn told the academy forum that the molecular biologists who first brought recombinant DNA technology to public attention "had no real precedents, no experience to go on . . . , no clearly visible apparatus to do the job better given that something had to be done at once." Maxine Singer and those who joined her in signing the letter from the 1973 Gordon Research Conference on Nucleic Acids to the president of the National Academy of Sciences unquestionably were on the side of the angels. They had no precedents, no experience, no apparatus. However, the events described earlier in this book show that the academy had the apparatus, the experience, and the precedents for guiding the Berg committee into paths that could have been as broad and well directed as those broken by the British Parliamentary Working Party headed by Lord Ashby and the implementing committee that bore the name of Sir Robert Williams. Indeed, Berg was told explicitly at the outset what the usual academy practice regarding selection of committee members was (chapter 12). Why that advice was later reversed by the academy president is one of the deepest mysteries of the long and bitter controversy over gene splicing.

I cannot be as charitable in judging the Berg committee's activities as Sonneborn has been, but I do agree with him that Berg and his colleagues acted sincerely in keeping with their concept of the public good. They were sure they knew what was best for the public. At certain times, in response to certain pressures, and to a limited degree, they acted in public view. But they did not at any time seek the public's concurrence in their decisions. They assumed that evidence acceptable to them would bring about concurrence. They were too preoccupied with their own opinions to recognize that their once automatic formula is not acceptable today. They were no longer preoccupied but arrogantly presumptuous when, backlashed by the effects of their original misjudgment, some of them sought to tip the scales by manipulating the evidence. On the assumption that it will be therapeutic for the scientists to read a straightforward set of statements of how their behavior of the last half dozen years looks to an informed layman who has great respect and high hopes for the scientific method but no inclination at all to look upon scientists as gods, I offer the following observations:

▸Recombinant DNA technology is just what those words say. It is a technology, a tool, or set of tools, that can be used in the conduct of basic research or in the development of a new form of mass manufacture of natural products. To control its use for the protection of the researchers and the public from harm does not constitute interference with anyone's rights under the First Amendment to the Constitution of the United States—Nobelist David Baltimore's emotional outcry against "the iron fist of orthodoxy" notwithstanding.

▸The deepest-seated fear that people have of gene splicing is that it will somehow be used to manipulate individual personalities to fit some arbitrary standard set by the manipulators at a time convenient to them. It is assumed that the subjects of the experiments would be powerless to defend themselves—children, old people in institutions and in hospitals, or members of some racial or political minority. The fear has little basis in fact at the present time because virtually nothing is known about putting together the genetic complexes that produce character, intelligence, or imagination. But it is conceivable that the distant future could bring justification for these fears, and it would be well for us, just to be on the safe side, to begin now to set up an early warning system against possible abuses.

At the National Academy of Sciences forum in March 1977, University of Chicago Philosophy Professor Stephen Toulmin compared these times of recombinant DNA technology's emergence with the days long ago when the advantages of fire were first discovered. Those days must have had their terrifying moments. But fire was too valuable a tool to taboo, and today we train professional firefighters who manage fire at its worst well enough for us not to think of abandoning fire as a comfort to our lives. Toulmin urged that we adopt the same attitude toward gene splicing. Just as we have arson laws to catch and punish incendiaries, so we must design genetic engineering control laws to catch and punish hybrid DNA molecule breeders who get out of hand. Then, having guarded ourselves against future surprise, we could feel free to take advantage of the contributions recombinant DNA technology almost certainly will make to the repair of human genetic defects and treatment or even prevention of hereditary diseases, which are already known to number more than a hundred.

We must feel free, for the blessings of gene splicing will come slowly. In the best mapped of all genomes—that of *E. coli* K12—the locations of only about 650 of the estimated total of 3,000 to 4,000 genes have been identified. In the human genome, Berg reported at the academy forum, a mere 150 genes have been mapped and no two of those are within two *E. coli* chromosome lengths of each other. That means there may be 5,000 to 10,000 unknown genes between any two known human genes.

The gene splicers themselves could accelerate growth of a sense of freedom from fear by dropping their insistence that recombinant DNA technology has no relationship to genetic engineering. No one believes that, and the longer they keep saying it the deeper will be the public suspicion that some hidden motive is involved—that science is withholding important information.

It is argued that anxieties about the future should be avoided. That argument would leave the future to chance. The future was left to chance at the time Henry Ford made the automobile a household commonplace. What have we now as a result? Fertile farmland paved over, closed to the planting of food crops. Rainfall runoff from the pavement periodically flooding sewage treatment systems and polluting streams we depend on for drinking water. The purity of the air we breathe polluted by automotive exhaust fumes. Sunlight acting on the fumes, generating smog and altering the climate around big cities. All of this ugliness might not have been foreseeable in Ford's day, but the logic of it is so straightforward that much of it surely would have been at least suspected if possible sequelae of the motorcar's advent had been considered then along with the pleasures of personalized transportation.

▸The measurable long-range value of gene splicing is necessarily debatable. But the immediate rewards for the involved scientists, in terms of their incomes and career advancement, are not. It is only natural for them to put their most optimistic foot forward (although, if they expect to maintain credibility in the public mind, they should present their claims in the normally accepted ways of science), but the rest of us would be irresponsibly gullible if we did not rigorously examine the reasonableness of their requests for our support. And, just as we are obligated to discount unduly rosy promises, so we should be skeptical of scare stories—undocumented accounts of the cloning of humans, for example—no matter how prestigious the publishers and distributors of the horror books. Above all, it should be kept in mind that, although there are theoretically sound reasons for believing that physically identical individual beings could be cloned when the detailed procedures were properly worked out, there is no scientifically accepted evidence that the minds and characters of the clones would be identical.

▸By predicting "breakthroughs" that are not likely to happen and repeatedly proclaiming their near approach when they are nowhere visible, the scientific community has skyrocketed public expectations not only of recombinant DNA technology but of all science and technology. This destructive charade could be ended if the word "breakthrough" were eliminated from the scientific vocabulary.

▸Up to now, most applications of recombinant DNA technology have involved a single strain of bacteria—*E. coli* K12. Although K12 had been a favorite organism for laboratory experimentation in genetics for many years before gene splicing was thought of, the Berg committee's failure to make an adequate search of the scientific literature (or even to conduct inexpensive exploratory experiments of its own, as was done by British scientists mentioned earlier), obscured a vital fact: that K12 does not thrive in the human gut nearly as well as do its multitude of wild relatives in the *E. coli* family or the other wild microorganisms that make up 99 percent of the bacterial population of the intestinal tract of warm-blooded animals. Consequently, until Roy Curtiss stumbled onto the truth while in the throes of creating χ1776, popular concern over the hazards of recombinant DNA technology was focused too narrowly on K12 to the neglect of other bacteria and viruses relevant to the technology. In their hurry to escape that spell of tunnel vision, the gene splicers (with the notable exception of Curtiss) are now generating an oppositely erroneous image in the public mind, one that suggests no reason at all for fearing unfavorable effects of the new technology.

At the academy forum of March 1977, genetic theory was expounded as an antidote to popular worries. Mutant strains of any species, went the argument, do not grow as fast, reproduce as rapidly, or compete for territorial space as aggressively as the main genetic line of the species does. As a generality, that argument cannot be faulted. Were matters otherwise, no species would remain distinct for very long. But that does not negate the cogent reasons that exist for exercising caution in speeding up the normal rate of genetic mutations in any species. It has been demonstrated conclusively—a fact we were again reminded of only recently by Professor Karl Z. Morgan, of the School of Nuclear Engineering at the Georgia Institute of Technology and a senior health official of the federal government for three decades—that *any* level of radioactivity has its degree of deleterious effects. Likewise, we may discover in time that any level of genetic mutation has consequences that do not always appear immediately. We just don't know, and we ought to be careful how we go about finding out. It is not necessary to look far in order to find impressive examples of the uncertainty of our present knowledge of how disease originates and spreads.

First the so-called Legionnaires' Disease. Legionnaires' Disease got its name by killing twenty-nine and sickening more than a hundred Legionnaires and members of their families who attended the fifty-eighth Annual Convention of the American Legion, Department of Pennsylvania, at Philadelphia from July 21 to July 24, 1976. It could just as well have been called the Magicians' Disease or the Candle-makers' Disease or the Eucharistic Congress Disease because it also attacked delegates to those conventions in Philadelphia during that summer. Although no one then realized that one disease agent was involved in all the cases, Legionnaires' Disease actually was called Broad Street Pneumonia because it also struck some people who were known to have walked on either one or the other or both sides of a certain stretch of Broad Street, one of the main thoroughfares of Philadelphia, during the Legion convention. On that stretch of street stood the famous old Bellevue-Stratford Hotel, one of America's most gracious stopping places for travelers, the site of the Legion meeting, and the momentary residence of many convention delegates. The Bellevue-Stratford ultimately had to close its doors and go out of business because the mysterious nature of Legionnaires' Disease could not be penetrated in time to reassure its accustomed clientele.

The symptoms of the disease were always the same—a tired, dispirited feeling, aching muscles, fever of 102 to 105 degrees, a generally dry cough, occasional vomiting and diarrhea, less occasional intestinal bleeding, and X rays that showed patches of lung congestion, which in the more serious cases consolidated into pneumonia. Oddly, the pulse rate, though rapid, was hardly ever as high as the degree of fever would normally call for. Typical death occurred seven days after the onset of symptoms; typical survivors left the hospital nine days after admission.

The epidemic first asserted itself on July 22, spread quickly for three days, caused its first death two days later, produced its worst death toll on August 1 (eighteen were dead by August 2), and petered out on August 16 without ever affecting residents of Philadelphia outside the Broad Street neighborhood centering on the Bellevue-Stratford. Legionnaires who left the convention feeling well came down with the disease after returning to their hometowns in widely separated parts of Pennsylvania but in no known case did the infection spread to other members of the conventioneers' families or to family friends or anyone else.

Because Edward Hoak, adjutant general of the Legion in Pennsylvania, set news hounds baying on the track of the mysterious killer on August 2—the day Dr. Sidney Franklin of the Veterans Administration Clinic in Philadelphia reported to the Center for Disease Control in Atlanta, Georgia, that eleven Legionnaires were dead of a feverish lung affliction that had also disabled forty other conventioneers—CDC staff doctors and nurses flew in from Atlanta; Washington, D.C.; Pittsburgh and Harrisburg, Pennsylvania; New Jersey; Maryland; Delaware; Ohio; Connecticut; and Florida to hunt down the unwelcome visitor. They located 182 victims (29 dead, 153 sick) ranging in age from three to eighty-two years but couldn't find the culprit despite tens of thousands of hours of searching by hundreds of members of the world's most expert team of disease detectives. Their professional reputations hinged on turning up a payoff for the largest investigation in CDC's history.

The clue hunt went to fantastic extremes: the number of pigeon feathers on the tops of air conditioners linking the hotel rooms to the air outside; contents of the dust on the ceilings of the elevators; sources of the ice cubes that were served with drinks in "hospitality rooms" maintained by candidates for office in the annual Legion election; components of insecticides sprayed on the trees that line Broad Street; causes of occasional wisps of smoke that rose for a few hours each day from a trash can several blocks away from the Bellevue-Stratford; even the identities of people in the Bellevue-Stratford lobby when sparklers were set off there while the Legion parade passed in front of the hotel.

Specimens of survivors' blood, urine, feces, and sputum and tissue samples from autopsies of the dead were rushed to the CDC laboratories for analysis. A hundred different scientific specialties and subspecialties were represented among those who performed tests designed to reveal the troublemaker. The intensity of the work was such that within forty-eight hours after receipt of the first specimens, infection from all known highly hazardous exotic agents was ruled out and within seventy-two hours the more common viral and bacterial diseases, including influenza, also were written off.

Tests for other known and unknown disease agents would take weeks or even months to complete. As they proceeded, telephone calls came into Atlanta from all directions. One was from the wife of a member of the International Order of Odd Fellows, who told of an outbreak of sickness during an IOOF convention in the Bellevue-Stratford in September 1974. CDC ordered a follow-up survey of some of the Odd Fellows who attended that meeting. A questionnaire brought 392 answers. Among them were eleven descriptions of symptoms consistent with a diagnosis of Legionnaires' Disease. Members of the lodge helped to identify and obtain clinical data on a total of twenty cases of the illness.

This unexpected break in the case, which occurred in October 1976, sent the CDC lab specialists back over the huge battery of tests they had made in August and September. And in December, Dr. Joseph McDade, who had responsibility for detecting rickettsiae (microorganisms that produce typhus among other diseases) in the original specimens and found none, reviewed his microscope slides in company with his chief, Dr. Charles Shepard. Neither man could see any rickettsiae. But both noticed the presence of a small bacterium with "suspicious consistency."

McDade had seen this organism before. It was in the tissues of guinea pigs that had been inoculated with postmortem lung tissue taken from victims of Legionnaires' disease in August. When cultures had been made of the guinea pig tissues, no disease-causing bacteria grew on them and the "suspicious" bacterium had been dismissed as one of the many inconsequential contaminants that are often encountered in testing for rickettsiae. The decision seemed to be the right one because pathologists had not found any pathologic changes in the Legionnaires' postmortem tissues that were compatible with bacterial pneumonia.

In reviewing the evidence, however, McDade and Shepard decided to try a different method of growing the "suspicious" bacterium from the tissues of the guinea pigs. This time they took samples from the spleens instead of from the lungs of the experimental animals. The spleen tissues had been held in deep-freeze storage. They were now thawed and injected into eggs containing chick embryos. When the embryos died five to seven days later, they were teeming with bacteria.

The scientists next took samples they had of serum from the blood of convalescent Legionnaires' Disease patients and mixed it with the bacteria removed from the chick embryos. Nine out of ten of the serum samples reacted with the bacteria, showing that the Legionnaires from whose bodies the serum had come had been attacked by the disease and had developed antibodies against it. The bacterium McDade and Shepard had discovered seemed indeed to be the cause of Legionnaires' Disease.

Why, then, had all the other investigators failed to implicate this bacterium? Further experiments demonstrated that the organism would not grow on any of the standard cultures in the time that was normal for bacteria—two days or less. So McDade and Shepard put a heavier than usual dose of the invisible bug on a particular type of growth medium and left it there for three days and sometimes longer. The bacterium finally grew and multiplied.

From this experience, CDC discovered the unusual nutritional requirements of the strange new bacterium. Continued examination of the organism did not reveal any characteristics similar to those of other bacteria, however. Nor did genetic analysis indicate any relationship between the bacterium and any other known bacterial species.

One more step had to be taken to clinch the identification of the cause of Legionnaires' Disease. The apparently guilty bacterium had to be shown to be present in diseased lung tissues obtained from autopsies of dead Legionnaires. All the widely used staining techniques failed to accomplish this task, which went unfinished until a seldom used stain called Dieterle's was tried. Dieterle's stain worked. The elusive bacterium was exposed at last. Its isolation was announced on January 18, 1977.

"All of the combined bacteriology and pathology experience accumulated since the beginning of the century pointed away from this agent being a bacterium," Dr. William H. Foege, CDC's director, testified at a regional hearing on Legionnaires' Disease conducted by Senator Kennedy in Atlanta in November 1977. "Drs. McDade and Shepard easily could have dismissed their observations as evidence of accidental contaminants. Fortunately, they did not."

After confirming that the Odd Fellows disease of 1974 was the same as the Legionnaires' Disease of 1976, CDC consulted its serum bank—an international repository for samples of blood from victims of many known and unknown diseases—and discovered that the identical microorganism had been responsible for a theretofore unexplained outbreak of sickness at Saint Elizabeth's Hospital in Washington, D.C., in July 1965. That episode was something of a local and regional sensation. It involved CDC in a frustrating investigation. Eighty-one people were stricken with "pneumonia" and fourteen of them died.

The CDC serum bank also disclosed that the still unnamed bacterial agent of Legionnaires' Disease had appeared in Michigan in July 1968, accounting for an outbreak of what was called Pontiac Fever. The bacterium attacked 144 employees of the Oakland County Health Department and visitors to the Health Department building, including several CDC officers assigned to investigate the epidemic.

There were no deaths and no pneumonia, but the symptoms were similar to the Legionnaires' Disease and there was no person-to-person spread of the malady.

In July 1973, a party of 252 Scottish vacationers traveled to Benidorm, Spain. They all stayed at one hotel. While on their holiday, ten of them caught pneumonia and three died. The governments of Spain and Scotland jointly investigated without coming up with an explanation for the tragedy. A blood specimen from one of the victims was preserved in Scotland, however, and after the cause of Legionnaires' Disease became known in 1977 a sample of that specimen was sent to CDC for examination. Antibodies against Legionnaires' Disease were taken from the sample. Later in 1977, another Scottish traveler went to the same Spanish town and stayed in the same hotel. He, too, died of pneumonia. Postmortem samples of his tissues and blood carried positive signs of Legionnaires' Disease.

Since August 1976, Foege told Senator Kennedy, sixty-four individual cases of Legionnaires' Disease, circumstantially unrelated to each other, had been seen in twenty-four of the fifty states. Sixteen of those people died. During 1977, clustered cases were confirmed in three states: nine in Ohio, with one death; twenty-one in Tennessee, with three deaths; and twenty-seven in Vermont, with fifteen deaths. CDC had begun formal training courses in laboratory diagnosis, and samples of the bacterium were being made available to qualified investigators upon request. "We assume," Foege said, "that the disease will be found in all fifty states, and perhaps in much, if not all, of the world."

A second example of the imperfect state of epidemiology comes from another witness at the Kennedy hearing on Legionnaires' Disease: Dr. Jay P. Sanford, dean of the School of Medicine of the Uniformed Services University of Health Sciences. Sanford testified that the rate of progress toward understanding Legionnaires' Disease had been "quite rapid, at least in comparison with many other infectious disease problems." He cited two earlier mystery diseases to document his statement.

One of these was Fort Bragg Fever, which probably appeared first in Wren, Georgia, in 1940, and was unmistakably present at Fort Bragg, North Carolina, in 1942, when forty cases were reported among army troops in training. Almost 10 percent of all officers and men in a given company were infected within two weeks. A commission of experts investigated. It blamed a virus, but could not identify the virus. Ten years later, a review of stored specimens showed that the cause of Fort Bragg Fever was *Leptospira autumnalis,* a bacterium that was known to cause disease elsewhere in the world with symptoms different from the symptoms observed in North Carolina.

Sanford's second mystery disease was the one described in chapter 2 of this book. Pneumonia broke out among army recruits at Camp Clairborne, Louisiana, in the winter of 1941–1942. A dozen years later, in 1955, the causative agent was identified as Type 4 Adenovirus. After passage of another decade, an effective vaccine against that virus was produced in 1965.

A third example of what epidemiologists don't know has to do with the flu. Influenza is the only remaining pandemic disease of humans, that is, a disease that rapidly spreads to large numbers of people in different parts of the world. A pandemic appears every fifteen years or so, bringing suffering to millions and death to thousands. Nothing can be done to stop a pandemic until the type of flu bug responsible for it is identified. One year we have Russian flu, another Victoria flu, another year swine flu, another year Hong Kong flu, another year Asian flu. Inoculation against one type does not guarantee protection against other types. Where do all the different types originate? Scientists have never been able to answer that question.

Recombinant DNA technology is beginning to provide some clues, however. The October 1977 issue of *Agricultural Research,* a technical bulletin distributed by the U.S. Agriculture Department, reported on experiments conducted jointly by Saint Jude's Children's Research Hospital at Memphis, Tennessee, and the Animal Disease Center of the Agricultural Research Service on Plum Island, off the coast of Long Island, New York.

In the first experiment, pigs and turkeys were caused to inhale three different viruses. When the animals and fowls were sacrificed later, the scientists found in their bodies new recombinant viruses different from any of the original three viruses. This proved that living creatures can create new viruses by recombining old viruses.

In the next experiment, one group of turkeys was infected with fowl plague virus. Another group of turkeys was infected with turkey influenza virus. Both groups of turkeys transmitted their infections naturally to other turkeys, and recombinant viruses appeared in the third group of turkeys. In fact, the recombinant viruses produced a mini-epidemic among all the turkeys.

In a third experiment, one pig in a herd of pigs was infected with the virus that caused the Hong Kong flu pandemic of 1968 in humans. Another pig in the herd was infected with pig flu virus. Within a week, other pigs in the herd were carrying not only the viruses they were originally infected with but recombinant viruses as well.

From those experiments, several other pieces of information about flu viruses were obtained. It was found that turkeys can have two different influenza infections at the same time, one in the upper respiratory tract and the other in the lower respiratory tract. The situation is perfect for facilitation of recombination of the viruses. Also, the dual infections can cause a "sparing" effect—the killing power of a highly virulent virus can be reduced or even eliminated altogether because of the opposition of a less virulent virus. And the recombinant viruses may become more numerous than the original viruses in infected birds.

Do these recombinant phenomena occur in nature? That is to be determined by future experiments. Whatever the truth may turn out to be, the Department of Agriculture is taking no chances. Its experiments have been conducted in the containment facilities on Plum Island, one of the safest places on earth for scientists to work with infectious viruses.

A fourth example of the danger implicit in relying too heavily on epidemiological formulas of the past is the new cancer virus that unexpectedly popped up in a laboratory in Texas in May 1976. Experimenters at the Southwest Foundation in San Antonio spliced together a harmless baboon virus and a harmless mouse tumor virus and produced a hybrid that caused tumors not only in mice but in dogs, chimpanzees, and baboons. It also grew well on laboratory cultures of human cells. Dr. S. S. Kalter, director of microbiology and infectious diseases at Southwest, told newsmen at the time: "When the chips came down, they scared us. I think our initial feeling was more of fear than anything else. My gosh, if it's doing this to the animals, what's it going to do to us? And then if it happens to us, it could happen to anybody."

Given the state of emotions that prevailed in much of the scientific community in the latter half of 1976, throughout 1977, and in the early months of 1978—which makes it normal either to withhold important information from the people and their elected and appointed representatives or to ballyhoo information that has not yet been confirmed—one can only wonder how many other examples similar to the fourth one cited here would come to light if the supposedly open public records really were open. Except for Berg's renunciation of the SV40-lambda phage *E. coli* experiment in 1972, the literature shows only one clearly hazardous effect of recombinant DNA technology that was cut off before it had an opportunity to express itself. Dr. A. M. Chakrabarty, a staff scientist at the General Electric Research and Development Center at Schenectady, New York, discovered how to make a hybrid bacterium that would digest fiber. In discussion with research colleagues, the vital function of fiber in maintaining a healthy gut was brought up. Chakrabarty saw at once that his ingenious new bug might produce debilitating or even fatal diarrhea, and he voluntarily abandoned his experiments.

▸Anyone who has studied science for very long knows that its fundamental shortcoming as an influence on everyday life lies in its necessity to chop problems up into small, well-defined pieces in order to be precise in examining each piece. The result is an immense amount of information about small pieces of sometimes very large problems. Unfortunately, when the small pieces are put together the behavior of the composite does not always correspond to what the behavior of an individual piece would lead one to expect. This is truer in biology than in any of the other sciences. It is true to a lesser but still great degree in physics when the laws that prevail in finite spaces are applied to such essentially boundless spaces as the oceans and the atmosphere (in the study of weather and climate, for instance) or, to go even farther out, the spherical shell of ozone that hangs in the upper atmosphere and protects all earthly life from being burned to a crisp by the sun.

This reality is pertinent to any intelligent discussion of the possibility of using recombinant DNA technology to multiply the earth's food supply by increasing the amount of nitrogen that can be drawn from the air to fertilize crops.

Nitrogen is a crucial element in our existence. It is the basic component in the A, C, G, and T of the genetic alphabet. Growing plants need nitrogen in order to continue growing. A few of them are able to draw nitrogen from the air with the help of bacteria that live in nodules on or near the plant roots. Centuries before the function of those nodules was understood, leguminous crops that bore them—alfalfa, for example—were rotated on a set schedule with other crops that did not bear nodules. When crop rotation proved too slow a method of restoring enough nitrogen to feed the roots of bumper plantings, artificial fertilizer was called into progressively more widespread use. The making of fertilizer today requires the diversion of natural gas from fulfillment of other needs of the country and ultimately creates a deficit in our international balance of payments.

Recombinant DNA technology may make it possible to breed hybrid bacteria or hybrid crops capable of drawing much more nitrogen from the air and thus reducing or eliminating the need for artificial fertilizer. But the hybrid bacteria would have to come from the soil-inhabiting families that Roy Curtiss mentioned in his testimony before the Stevenson subcommittee. There are potential hazards in the use of those bacteria, for the reasons Curtiss gave. It is not reassuring to learn that the bacterium now known to cause

Legionnaires' Disease is thought by some scientists to be a soil resident. And if, in spite of these problems, it proves possible to draw more nitrogen from the air, we may have to be careful in drawing it so as to avoid a natural catastrophe that would do more damage than the nitrogen could do good. Too much free nitrogen, some scientists say, could deplete the ozone layer and expose us to the incinerating effects of ultraviolet light from the sun. Or, if the freed nitrogen were absorbed by the oceans, the water might react by trapping carbon dioxide to contain the resulting ammonia. This would lower temperatures on the planet (which are regulated by carbon dioxide in the air), defeating the original purpose of recombinant DNA technology by shortening the growing season.

▸The uproar over recombinant DNA technology control legislation that has been going on in Congress for the last four years began because the NIH guidelines covered only gene-splicing experiments financed by public monies. Privately owned industry was free to do as it liked, which is to say free to do whatever seems profitable. From this circumstance has arisen the impression that owners of private businesses—drug makers especially—are responsible for stalling the control bills in Congress, and have meanwhile been racing pell-mell to gather as much profit as possible before a law is enacted to slow them down.

Anyone who has attended the congressional hearings on recombinant DNA technology knows that this prevalent impression is utterly false. Only a handful of private companies are actively engaged in gene splicing at any level. Being accustomed to dealing with pathogenic organisms routinely, they have a very realistic idea of what the dangers are and how disastrous damage suits brought by victims of accidents can be. Private industrialists actually are pushing for a law, one that provides for licensing, inspections, and appropriate penalties for violations. They need a solid footing from which to calculate their insurance premiums, which are a normal part of any company's operating costs.

Gene-splicing scientists in the universities are the ones who have been stalling the bills in Congress and (this would be amusing if it were not also rather sordid) a number of them are patent seekers and part owners of companies, which smart stockbrokers showed them how to set up for later sale. The ultimate buyers—after the process of going public brings in enough investors to recoup the scientists' nest egg at several times its original size—will be established private firms who have earned or can afford to borrow the generous funds required to pay their way through the long period of development that lies ahead for recombinant DNA technology's products. When those products finally reach the marketable stage, they will, of course, be subject to the controls of the U.S. Food and Drug Administration and other regulatory agencies. Only then will we really see how well the involved scientists have kept their promises to bring consumers better medicines at lower cost.

▸As anthropologist Margaret Mead said in her brief testimony before the Stevenson subcommittee of the Senate in November 1977, there never was and is not now any reason for unduly hurrying development of recombinant DNA technology. Regardless of past propaganda and some present appearances, results of gene-splicing experiments that can be measured in the practical terms of everyday living will be a long while coming.

The NIH's formal statement of the impact of recombinant DNA research on the environment, issued in October 1977, said flatly that the "expected practical applications" of DNA molecule hybridization "have not yet been realized and their success remains uncertain . . . at this time." Only two higher eukaryote genes had been reproduced by the prokaryote *E. coli* K12: the gene that controls production of insulin in the pancreas of the rat and the gene that controls production of human globin, the protein part of hemoglobin, the carrier of oxygen in the red blood cells from the lungs to tissues throughout the body. In neither case was the protein called for by the involved gene replicated.

The NIH environmental impact statement reduced the following to the status of "prime candidates" for future development: human insulin, human growth hormone, blood-clotting factors, specific antibodies and antigens for the prevention and treatment of specific diseases, and enzymes useful in the management of embolism. Application of gene-splicing technology to energy production through manipulation of the genes of photosynthetic organisms and to neutralization of environmental pollutants such as oil spills at sea were relegated to a yet more "highly speculative realm."*

Since publication of the environmental impact statement, the gene coding for the human brain hormone somatostatin has been replicated and the hormone itself has been reproduced by the χ1776 bacterium. Although, as we saw earlier, academy president Handler used his position to advertise the somatostatin exploit in a conven-

* The United States Court of Patent Appeals early in 1978 granted to General Electric a patent on a hybrid strain of *Pseudomonas* bacterium that degrades crude oil more completely than does any bacterium found in nature. But GE has made no prediction about how soon the patented bacterium will be ready for marketing.

tionally forbidden way, he attached a weighty caveat to his unorthodox sales pitch:

> From this historic point [marking the first expression of a human gene in a bacterium] to the practical production of insulin, or growth hormone, or any of the enzymes that might be useful for repair of genetic disease, or to production of antibodies and clotting factors, is still a very long distance. These . . . proteins . . . are much more complicated [than the string of fourteen amino acids that make up somatostatin or the eighteen aminos that constitute its chemical precursor, somatotropin . . . and for them] the simple spinning out of the strand of amino acids is insufficient. Each requires additional processing by various kinds of enzymatic procedures which may not be normal *E. coli*. And it may be an even farther distance to the useful incorporation, by this process, of nitrogenase [the enzyme that draws nitrogen from the air to fertilize legumes through nodules growing on their roots] into wheat or corn. A great deal remains to be done before there can be a true practical payoff.

All that being so, we can afford to mark time while we test experimentally—instead of continuing to guess at—the degree of risk that resides in various types of recombinant DNA technology. Those risk assessments were supposed to have been done first, before there was such a thing as a guideline. But four years have passed between the time the NIH responded to the Berg committee letter and the beginning of April 1978, when the first risk assessment experiment was started by cloning a hybrid cancer-causing virus in a bacterium and then seeing whether the cancer could be reproduced in mice in a one-room P4 laboratory hastily constructed inside one of the buildings of the U.S. Army's abandoned biological warfare compound at Fort Detrick outside Frederick, Maryland. There had been much reluctance to undertake the risk assessments, the grounds being that the results might be equivocal and therefore valuable ammunition for the critics of gene splicing. Once British researchers began similar experiments in a former biological warfare lab in England, however, the American foot-draggers were too embarrassed to continue their delaying tactics.

A complicating factor in the risk assessment impasse was the time NIH wasted in its attempts to evade its legal obligation to respect the National Environment Protection Act of 1970. Congress passed NEPA as a means of requiring executive departments and agencies of the government to offset their steady flow of favorable advertising of themselves and their work with occasional declarations of their weaknesses and the problems their activities sometimes generate.

These declarations are the environmental impact statements (EIS). When NIH failed to file an EIS with its first set of guidelines in 1976, Frederick J. Mack, Sr., an attorney who lives in Frederick, near the old Fort Detrick P4 facility, instituted a lawsuit to halt renovation of the P4. His reason was not that P4 containment would be inadequate but that sloppy experimental conditions allowed by the NIH guidelines would endanger the town where he lived. NIH belatedly filed an EIS in the autumn of 1977 but then proposed guideline amendments without submitting an EIS to cover them. Mack tried to reopen his suit but the courts would not accept the maneuver. More recently NIH has supported legislation designed to remove NEPA's requirements where gene-splicing guideline amendments were concerned. Environmentalists found this special pleading galling. For if the rationale for the guideline changes were as persuasive as NIH claimed, the writing of a new EIS should be hardly more than a simple exercise in penmanship.

▸There was reason to suppose that a truly honest EIS could not tell so simple a story. Only a fortnight after the unhappy meeting of the Advisory Committee to the NIH Director in mid-December 1977, *Nature* reported nervousness in the British research community over the extreme liberality of the proposed revisions of the NIH guidelines. The message was clear: "Many British scientists feel that, in the present state of knowledge," the proposed guideline revisions "may be relaxing things a little too far."

The most worrisome problem, the *Nature* correspondent said, lay in the suggested guideline change that would allow physical containment requirements for experiments to be reduced in proportion to the rise in efficiency of biological containment. Although χ1776 had been approved for some high-risk experiments in England, and Britain's "disarmed bug" pioneer Sidney Brenner was reported to be almost ready with a rival for χ1776 (and Curtiss was about to report on a second generation of χ1776), the British system of recombinant DNA technology control still put much more reliance on physical restrictions than the NIH guidelines did. As physical containment is more expensive than biological containment, the report in *Nature* observed, adoption of the NIH guideline changes could result in a "serious discrepancy between British and American laboratory practices . . . with potentially embarrassing consequences. It could, for example, be cheaper to send a research worker to the U.S. to carry out a set of experiments requiring little more than a conventionally equipped laboratory bench than to install—or even rent—facilities providing the higher physical containment levels required to carry out the same experiment in the U.K.

"Such embarrassment is likely to be reinforced if other European countries, most of which have so far followed the British guidelines ... decide to break ranks and, in line with the U.S., introduce significantly lower containment levels (as the French are, indeed, now proposing to do).

"If, for example, Germany decides to introduce less stringent guidelines, British scientists working at the European Molecular Biology Laboratory at Heidelberg, which operates under German laws, may find they are permitted to carry out experiments which they are unable to do at home."

The effects of what happens in Heidelberg are felt all over Europe and indeed around the world. For the European Molecular Biology Laboratory is run by the European Molecular Biology Organization (EMBO), a private, technically oriented grouping of representatives of seventeen nations: Austria, Belgium, Denmark, Finland, France, German Federal Republic, Greece, Iceland, Ireland, Israel, Italy, Netherlands, Norway, Spain, Sweden, Switzerland, and the United Kingdom. The EMBO laboratory at Heidelberg is shared by scientists from nine western nations and Israel. The director of the lab is Sir John Kendrew, who is also executive secretary of the International Council of Scientific Unions (ICSU), whose eighteen unions and sixty National Academies of Science and research councils include East European and Third World countries as well as the U.S.S.R.

ICSU in 1976 created within itself a Committee on Genetic Engineering (COGENE), which recently conducted a worldwide survey that showed sixteen countries with recombinant DNA research guidelines: Australia, Belgium, Canada, Denmark, England, France, German Federal Republic, Ireland, Israel, Italy, Japan, Netherlands, Norway, Switzerland, and the U.S.S.R. Aside from its ties with ICSU through Sir John, EMBO maintains close liaison with the European Science Foundation (ESF), a nongovernmental aggregation of forty-five national research committees from sixteen nations (subtract Finland from EMBO's list of members and you have the ESF membership) headquartered at Strasbourg. ESF is preparing to advise its members on registry, licensure, inspection, and record keeping at recombinant DNA research sites. It has a European Liaison Committee for Recombinant DNA Research on which the European Medical Research Councils (EMRC) are represented (to get the membership of EMRC, subtract Greece, Ireland, and Israel from the EMBO list). EMRC favors registry and uniform regulation of all recombinant DNA activities.

This nongovernmental but nonetheless politically powerful scientific complex inevitably influences thought and action within the

European Economic Community (EEC)—whose European Commission has been considering issuance of a directive requiring registry and advance approval of recombinant DNA experiments by appropriate national commissions—and also within the United Nations (which has a particular interest in nitrogen fixation experiments as a means of enhancing the food supply in underdeveloped countries and in training microbiologists to meet the exacting requirements of recombinant DNA work) and the World Health Organization (WHO). One of WHO's primary concerns is to guarantee safe dispatch, transmission, and receipt of infective organisms.

Although European laboratories late in 1977 had only 150 recombinant DNA experiments under way, compared to 300 in the United States (while Australia, Canada, Japan, and the U.S.S.R. together had 20 to 25), anticipation of global traffic in recombinant DNA technology products was apparent in the work that the World Intellectual Property Organization (WIPO) had been doing quietly behind the scenes since 1973. The activity had originated from the fact that disclosure of an invention is a generally recognized requirement for a grant of patent. Normally, an invention is disclosed by means of a written description. But where the invention involves a microorganism or use of a microorganism the written description is not enough. A sample of the microorganism must be deposited somewhere for inspection.

Patent offices are not equipped to maintain repositories of this kind because of the special requirements of safety and protection against contamination. So, in 1973, the year of the first Cohen-Boyer experiments with DNA molecule hybrids, the British proposed that WIPO study the possibilities of establishing a single repository to serve all the peoples of earth. That would be a sizable undertaking, for the microorganisms would have to be kept intact for at least thirty years to cover the span of patent duration. The executive committee of the Paris Union for the Protection of Industrial Property approved the British initiative, and a committee of WIPO experts threshed out details in 1974, 1975, and 1976. The need for a formal international document was recognized, and during four days of April 1977, a treaty was drawn up in Budapest, Hungary.

Considering all this, it was only natural that the British would be concerned about the consequences of the NIH proposal to revise the guidelines for recombinant DNA research in the United States. NIH had not, after all, obtained the approval of Congress in any form, even for the original guidelines, let alone the amendments. The British, on the other hand, had been punctiliously correct about their approach to the control problem from the beginning. The British Parliament had been first to give official sanction to gene splicing under appropriate legal control, and the control system that the British devised was far more expressive of democracy in action than was the patchwork system that NIH later produced. Lord Ashby's Parliamentary Working Party had laid out the basic design in January 1975, and the Williams Committee had followed up in 1976 with provision for a Genetic Manipulation Advisory Group (GMAG) of nineteen members headed by Sir Gordon Wolstenholm, director of the CIBA Foundation, and including others from industry along with representatives of labor unions and the public as well as the scientific community. GMAG was given authority, as agent for the Secretary of State for Education and Welfare, to decide in advance, with the advice of the Medical Research Council, what recombinant DNA experiments would be allowable at what levels of containment. The decisions were put on an individual footing, so as to gradually build up a new branch of case law. Privately financed as well as publicly funded research was to be covered.

If the NIH guideline revisions were to create the discrepancy between British and American laboratory practice that *Nature* foresaw, the consequences, the journal pointed out in an understated way, would extend beyond the material plane and penetrate the very spirit of democracy. As the *Nature* correspondent put it:

> GMAG could find itself caught in a . . . difficult dilemma. For while responding . . . to demands from scientists to rationalize containment levels is [reasonable up to a point] public confidence requires that it [GMAG] must not seem to be responding to outside, and particularly industrial, pressure. The members of GMAG are conscious that if they impose too harsh a set of restrictions on industrial research programs as compared with other countries, companies will merely transfer their research programs—and their revenue-producing potential—elsewhere. . . . However, the Williams Report . . . is still less than two years old. Any major attempt to re-draft its proposals at this stage is likely to be seen as premature, if not by scientists at least by the public, to whom some of the more fanciful possibilities (and dangers) of genetic engineering have only recently come home. . . .
>
> It is one of GMAG's strengths that, through the trade union and lay representatives, it is able to take public concern directly into consideration (thus avoiding the major confrontations between scientists and the public that have, for example, occurred in the U.S.). Yet, as the pressures on GMAG increase, so the tensions between its constituent parts may become more difficult to control. Such tensions arise partly from the ambiguous nature of GMAG itself, seen by some as an unfortunate, if necessary, part of scientific bureaucracy, and by others—

including the trade unionists—as a successful model of public participation in research policy, with potential applications in other areas.

Two weeks and a day after publication of that muted London signal of trouble on the horizon, members of the 95th Congress returned to Washington, D.C., for the opening of their second session on January 20, 1978, and immediately became aware of Harvard University's demand that any recombinant DNA legislation include preemption by the federal government of state and local authority to intervene in the writing of guidelines.

Testimony on the desirability of preemption had been heard at the Stevenson subcommittee hearings of November 1977. "We have a federal system of government," said Assistant Professor David Newberger of Washington University Law School at St. Louis—a student of business regulation specializing in the impact of regulation on innovation in American society. "Under a federal system of government, the states have a right to protect the health and safety of their citizens, just as the federal government exercises some rights in that area on a nationwide basis. . . . Unless there is some national purpose in preemption, then we ought not to be doing it."

No national purpose would be served by capitulating to Harvard. The capital looked to Senator Kennedy for a sign of his intentions. Would he sit silent if Congress acted to skew the nation's perspective on one of its fundamental political tenets and pull the country back from its rightful role of leadership in the shaping of a historic new phase in human affairs? His chief lieutenant in the recombinant DNA wars, Stanley Cohen's "old friend Larry Horowitz," ducked all queries.

The timing was not helpful to American prestige. For with the coming of October 1978 the United Kingdom's Health and Safety Executive (HSE) would assume responsibility for enforcing the British GMAG standards under the Health and Safety at Work Act. That law authorizes HSE to monitor the condition of workplaces, conduct inspections, and impose penalties where deserved. Well before then, the House of Commons Select Committee on Science and Technology would open public hearings on the future of genetic engineering.

Already some within the scientific community were bewailing the possibility of "anti-scientific feeling" among the MPs. But that timeworn cliché was not being heard in the customary vacuum. The newly designated Genetic Engineering Group of the British Society for Social Responsibility in Science had assured a balanced dialogue by announcing its approval of the hearings and declaring its intention to testify in opposition to the "complacency" it found among scientists on matters of safety.

It was not only in England that the wishes of the people were being taken more seriously into account than they were in the United States. The Netherlands—whose representative at an International Council of Scientific Unions meeting in Washington, D.C., in October 1976 had sharply rebuked U.S. Academy of Sciences President Handler for insisting on freedom to pursue knowledge of every sort and kind ("there is some knowledge we are better off without," the blunt Dutchman had said, citing examples of learning that could be acquired only through cruelty to people or to animals)—was prepared to establish control over genetic engineering as a whole rather than over gene-splicing alone, to have all experiments registered under the joint supervision of agents of government and society as well as of science, to monitor compliance with certificates of public trust, and to punish violators. Japan was agreed on registry of all recombinant DNA experiments during the planning stages to protect both experimenters and the public, on subsidy of safety equipment by the government, and on creation of a special government-supported educational system that would include overseas training in safety techniques. Even the usually unflappable Swedes, whose government had shown no interest in recombinant DNA technology in 1973 (sending risky experiments to the Pasteur Institute in Paris rather than bothering with them at home), in 1978 were arguing whether local authorities as well as the national government should have a say in the regulations. The immediate focus of attention in that quarter was Uppsala University's asserted need for a P3 facility by 1979 to deal with *B. subtilis,* the favored (but not yet approved) alternate to *E. coli* as an experimental organism.

The unaccustomed attention to laboratory safety everywhere was noted with approval in a public paper distributed by WHO. It called the new trend "one useful outcome of the recombinant DNA controversy." The same document carried a piece of advice on legislation pending in the various legislatures of the world: "responsibility for damage to people and property should be clearly attributable."

All those and other details of the global situation were monitored and quietly published by an international subcommittee of the Interagency Committee that NIH Director Fredrickson established in 1976 in response to a directive from the White House. The Advisory Committee to the NIH Director presumably was acquainted with those details. But neither Fredrickson nor any of the committee members apparently saw in them the rare opportunity they provided for a dramatic lift in the level of American involvement in the plainly expanding new phase of participatory democracy.

The Advisory Committee to the NIH Director had never been encouraged to act as a committee in a truly advisory sense. Instead, it had been treated as a grab bag of prestigious citizens, each of whom is given a few minutes to state his or her individual views and later to write the NIH Director a letter confirming or expatiating on those views. This practice was an underlying cause of the displeasure expressed during the committee meeting of mid-December 1977, when discussion of the proposed amendments to the recombinant DNA research guidelines was railroaded in such an obviously superficial fashion. Fredrickson could have restored some of the dignity he lost in the Stevenson subcommittee hearings of November 1977 if he had, in the following month, summoned the Advisory Committee to conduct a formal hearing of witnesses subject to direct questioning and cross-examination. After adjournment of such a proceeding, the committee could have reassembled to draft a considered statement of its opinion and finally published majority and minority views.

Fredrickson could have done that, with great credit to himself and even greater reassurance to an apprehensive public. He could have, but he did not. Bystanders could draw only one conclusion: he wasn't interested in having by his side a public watchdog with a strong set of teeth.

▸The crucial unanswered question about gene splicing is not whether hybrid DNA molecules can be contained within the scientific laboratories but whether the scientists who breed the hybrids can contain their own desire for individual freedom enough to save themselves from destroying the vital bonds of mutual faith and trust beteeen science and the people who pay the bills for scientific experiments. Throughout the year 1977 and the spring of 1978 the repeated cry from the biology labs of academe had been, "Let us alone! We know what we are doing! Our procedures are entirely safe! We are quite capable of governing ourselves in a responsible manner without being policed by political hacks!" Unfortunately, the gene splicers' behavior fell far short of their promises. What happened at San Francisco was more than dismaying. It was frightening. The record of events there, as pieced together by the Stevenson subcommittee of the U.S. Senate, could be inserted into the scenario for *The Godfather* without seeming at all out of place except for the absence of physical violence.

The testimony quoted earlier in this book is a sufficiently sad account of disrespect for the democratic process. But the letters, memos, and other documents introduced into the official transcript during the four months following the end of the hearings show that virtually no one told the whole truth on the witness stand.

Professor William Rutter's original story was that he, on his own initiative, had ordered destruction of ten rat insulin clones that were the only hard evidence of violation of the gene splicing guidelines. He did not explain that before the clones were destroyed they were analyzed, that six of them were found to be recombinant DNA molecules, and that one of the six was specifically coded for insulin. Having obtained that information by illicit means, the scientists involved knew that when they could repeat the experiment with a less versatile but legal cloning organism, they could count on getting the result they were looking for. Contrary to the report published in *Science* in September 1977, the guideline violation therefore did give the San Francisco group an unfair advantage over rival experimenters.

Worse still, Rutter failed to mention in his testimony that before doing anything with the clones he telephoned an NIH official in Washington and asked that official's "informal" advice on what to do. The call did not go to Dr. William Gartland, head of the NIH Office of Recombinant DNA Affairs, the responsible person who should have been notified. Instead, Rutter called NIH Recombinant DNA Molecule Program Advisory Committee Chairman Dewitt Stetten. It was Stetten who advised disposing of the clones. When and under what circumstances the disposal was to occur are still not defined in the Senate hearing record, and if Stetten advised anyone else in NIH about his conversation with Rutter, the record does not show that, either.

When UCSF Biosafety Committee Chairman Martin first got wind of the violation several months after Rutter's call to Stetten, no immediate report was made to the whole Biosafety Committee. When Martin did report, on June 3, he asked the committee to approve a letter addressed not to NIH ORDA Director Gartland but to Stetten, complaining in general terms about communications difficulties between UCSF and ORDA but not mentioning the guidelines violation. "Obviously, this is a time when communication should be most effective in order to avoid unwarranted problems and anxieties among both scientists and lay members of society," Martin wrote. "The climate is right for development of a hysterical reaction stimulated by miscommunication. . . . I hope you can find the time to look into what appears to be a most serious deficiency in that office [ORDA]. If you would like any further documentation of the problem . . . we shall be happy to provide it."

Stetten didn't need further documentation; he had known of the guideline violation for weeks. But he did not refer to that awareness in his reply to Martin, which thanked "Dear Dave" for "reviewing what certainly appears to be a collapse of appropriate communications between our Office of Recombinant DNA Affairs and workers at the University of California, San Francisco." After offering ORDA's understaffing as justification for the problems, Stetten concluded, "All this is but a poor excuse for the maltreatment to which you have been subjected. I shall certainly raise the issue with Bill Gartland, who is both intelligent and industrious, and I am certain he will do everything in his power to rectify the present difficulty."

Whatever else he may have told Gartland, Stetten did not disclose to him that a guideline violation had been committed at San Francisco, and when *Science* finally broke the story in September 1977, Gartland, taken completely by surprise, became the scapegoat for the whole matter: NIH Director Fredrickson did nothing to defend his ORDA chief before the Stevenson subcommittee, and did not even hint at Stetten's curious role, which was likewise passed over by Senators Stevenson and Schmitt.

Professor Herbert Boyer in his testimony had pleaded ignorance of the guideline violation at the time it occurred because the experiment in question had not been done in his laboratory. According to one item in the hearing record, he had actually been designated as the responsible alternate for the head of the lab where the experiment was performed, and NIH had accepted his designation. But Boyer continued to insist that he "did not proceed without confirmation of the approval of the pBR322 plasmid . . . was in no way involved with overseeing the activities in Dr. Goodman's laboratory . . . and was not aware of the progress of the rat insulin gene cloning experiments until March 1977."

On February 4, 1977 Boyer had called a meeting of laboratory workers to tell them the plasmid pBR322 had not yet been approved as safe for use. At that time, of course, pBR322 had already been used in that very lab. Did anyone at the meeting tell Boyer so? The hearing record indicates not. Why not? No one seems to have asked that question.

The record shows very plainly, however, that at least one scientist at the UCSF Medical School knew of pBR322's unapproved status at the time pBR322 was used in the rat insulin gene cloning experiment which began on January 16, 1977. He was Dr. John Baxter, a member of two departments at UCSF: the Department of Biochemistry and Biophysics, within which the guideline violation occurred, and the Department of Medicine. Sometime in January, in a grant application to NIH ORDA, Baxter stated his understanding that pBR322 had not yet been approved and he agreed not to use it in the hormone experiments he was working on.

How could Baxter have been so sure when so many others pleaded either ignorance or confusion? It was a very good question. Like many other good questions, it went unanswered.

The simple truth (which had to be read between the lines of the testimony) was that the scientists in the Biochemistry and Biophysics Department were playing the game of the old school tie. Their system of communication was built on personal relationships and was used as often to avoid official channels as it was to respect them. The purpose was to get "inside" information before it became generally available and thus to maintain the authority of the tightly knit oligarchy that has run science counter to the democratic process. The system breaks down further with every step toward public participation in policy decisions concerning science, which is an important reason why many now powerful scientists don't want public participation. In this particular instance, the inside "tip" did not take into account NIH Director Fredrickson's eminently sensible insistence that he have data before him in writing in order to study it carefully before approving or disapproving the decisions of subordinates. Laboratory confirmation of certain tests of pBR322 (tests which had been requested by qualified scientists) had been given by telephone (in oral form, they could be interpreted or even revised later if circumstances suggested such a course) but had not been set down in less retractable black and white.

In spite of the repeated suggestions of duplicity and deception, Rutter and Boyer persisted in claiming that the gene splicers were behaving honorably and responsibly. After seeing gally proofs of the Stevenson hearing transcript, they wrote to the Senator complaining bitterly about the quizzing they had got on the witness stand. But their unhappiness was as nothing compared to the Olympian indignation of Dr. Waclaw Szybalski, of the University of Wisconsin, a member of the Recombinant DNA Molecule Program Advisory Committee of NIH. He swept aside the fact that neither Rutter nor Boyer had claimed to be experimenting with anything more advanced than the insulin genes of the rat and had not suggested that even those genes had been expressed in the manufacture of rat insulin. Szybalski rebuked Senator Stevenson for holding up progress toward the production of *human* insulin. Szybalski said the Senator was derelict in failing to "commend Drs. Rutter, Goodman, and Boyer" for using pBR322; indeed, Stevenson should have scolded the scientists for not continuing to use pBR322 prior to its official approval. Szybalski asked, "How could a scientist, who knows that he might be able to save many human lives and prevent human misery, delay his research just to avoid some unjustified criticism?" The choice, stated in Szybalski's terms, was between "following scientific reason and . . . conscience" or "just satisfy [ing] out-of-date bureaucratic regulations." In abandoning pBR322, he said, Rutter and Boyer probably contributed to this country's "rapid loss of leadership" in gene splicing "to European countries, including the USSR"—a loss that no one but Szybalski had noticed.

Senator Stevenson's reply was unequivocal, "I remain very concerned over the conscious decision of Dr. Rutter and his associates not to report to NIH the violation which occurred when there was no longer any doubt that they had not followed, however inadvertently, the established procedures. . . . The scientific community has the responsibility to follow whatever guidelines are in effect until the rules are changed in an appropriate fashion. I am frankly surprised at your suggestion that Senators should commend scientists for violating the regulations simply because the scientists were engaged in important and useful work."

Scientists are a persistent lot, and I am sure that some of them will say I do wrong in emphasizing one incident at one laboratory. They might have a case if the San Francisco episode were unique. But there have been other instances of scientists taking policy into their own hands. Aside from the case of Professor Charles Thomas at Harvard, which has great significance because Thomas—like Szybalski—helped write the NIH guidelines, there have been various revelations of a widespread conviction among scientists that they have a right to decide for themselves what they should be doing with public monies and how they should be doing it. The Stevenson hearing transcript includes a quotation from a letter one biologist wrote to the Environmental Protection Agency, "The whole question of regulation and monitoring is abhorrent, especially when done by a government agency and not by scientists." A Yale University scientist in a letter to Stevenson said, "I was present at a conversation in which a well-known scientist, when it was pointed out to him that an experiment that he was in the process of doing violated the guidelines, replied, 'Oh, well, the adverse publicity will just give the work greater renown.' Obviously, he did not feel threatened by the possibility of NIH-enforced sanctions."

I believe I understand the workings of the mental trap in which too many (but far from all) scientists have imprisoned themselves. The mechanism is triggered by the fact that the results of no worthwhile experiment can be predicted in advance. The competent scientist starts out to test certain hunches. As soon as he is persuaded that the original hunches were wrong, he shifts to a line of inquiry that looks to be more productive. He may consult with his immediate colleagues before making the shift, but he relies most strongly on his own judgment. In purely technical matters there is no escape from this system. The trouble occurs at the point where the technique assumes a high probability of affecting the health or safety of lab workers or of people beyond the lab.

The wise solution to the problem is precisely as Dr. Robert Pollack outlined it at a public hearing on gene splicing conducted by the office of the Attorney General of New York State. Because of his early influence in the gene-splicing field, Pollack had been named to a biohazard committee at the Stony Brook campus of the State University of New York. He found his research colleagues reluctant to respond to his efforts at being a friendly policeman. He was embarrassed. He was wasting time in petty contention—time that should be spent on his own experiments. "I do not wish to see the period of self-enforcement prolonged," he testified. "I have come to the conclusion that the guidelines should have the force of law."

Before I end this book, I must complete my identification of myself and my motives by reporting a challenge that has been raised by one of the scientists whose behavior I have criticized. Responding to some nettlesome questions I could not escape asking him, he said, "John, you told me you were going to report what has happened in the recombinant DNA controversy. But you are not doing a straight reporting job. You are working on a thesis."

True. I do have a thesis. It came from the most influential amateur geneticist who ever walked the earth. An extremely unusual man, he lived a short life about 2,000 years ago and died a dreadful death because most of those he lived among could not understand what he was saying—that although he was born of an earthly woman his parentage extended far beyond the earth. He personalized the story to entice a primitive audience, but it paralleled the story told in the prologue of this book to explain the earthly presence of all humans.

The man called Jesus taught that a primary rule of inheritance grants to each of us a set of talents which we are expected to use for the common good. If we don't use them, he said, they will wither away; they will be of no use even to ourselves if we try to hoard them.

Hidden in this philosophy are some implications not at all consonant with modern American practice of democracy. One is that all of us humans are not created equal, despite the mighty efforts of the fathers of our country to make us so. We are not universally interchangeable parts capable of dropping ourselves at will into any assignment that seems desirable or convenient. By ignoring this truth that every biologist knows, the launchers of recombinant DNA technology generated some critical problems that sooner or later will have to be faced and resolved.

The reasonable approach would have been to limit the use of this exquisitely versatile, powerful, and potentially dangerous tool of genetic manipulation to those who were capable of understanding it and trained to use it. That was what Sinsheimer and others got hooted at for proposing. The hoots masked the true nature of the opposite tack that was being taken. Every possessor of an academic degree in science or a grant of a fellowship to try for a degree was to have a crack at Pandora's box.

Are we foolish enough to allow gene splicing to deteriorate into a high school laboratory stunt, prey to the whim of any adolescent? I see one strong hope for a negative answer. It lies in the influence of American motherhood over the well-being of its children. The opponents of runaway gene splicing have failed up to now because they aimed at the wrong target. They tried to stop recombinant DNA technology altogether. It wasn't a reasonable objective. The failure is now understood and the focus of attention has been narrowed to attainment of public participation in the shaping of science policy. At the center of the regrouped forces is a core of undeniably intelligent, tough-minded women: Pamela Lippe, of Friends of the Earth; Deborah Feinberg, of the New York State Attorney General's office; Professor Dorothy Nelkin, of Cornell University; Susan Wright, University of Michigan lecturer in the history of technology; Marcia Cleveland, legal counsel for the Natural Resources Defense Council; and last but very far from least, Dr. Margaret Mead, emeritus curator of the American Museum of Natural History.

Against that stalwart array it will be extremely difficult for the NIH and the Congress to continue to parrot the old cliché that the people aren't prepared to make intelligent decisions. It never was a sound argument, of course. If it had been (it is interesting that such an experienced science hand as Sonneborn didn't think it was), then the scientists were to blame for a failure in teaching. Instead of accepting that responsibility and moving to correct the deficiency, however, those scientists who complained the most bitterly threatened to leave the country or to find other employment for their talents. No one believed that the threats would be carried out, for there was no more generous country for the scientists to go to, no better life than the one they enjoy here. Nevertheless, the fantasy continued to be projected, and President Carter did nothing to dispel it in his 1978 State of the Union message to Congress.

A moment when careless or non-existent recordkeeping was being advanced as an explanation for violation of government regulations was not an appropriate moment for the President to speak of paperwork as an oppressive chore and to promise relief from its onerous influence. Nor was a time of scandalous lack of accountability for expenditure of public funds on scientific research an appropriate time for the President's science advisor to express willingness to lengthen the periods covered by government research grants. Events were clearly declaring an urgent need for rigorous discipline, but the image the White House presented to the recalcitrant scientists was that of forgiveness for past transgressions and eagerness to provide larger opportunities for repetition.

It was never too late, of course, for the President to intercede by acting on a suggestion that William Carey, executive officer of the AAAS, had repeatedly made during his days at the White House—that an annual Presidential report be issued to advise the Congress and the people of the substance, organization, costs, goals, problems, and progress of science and technology during the year. Or, if more convenient, it certainly would be timely for the President to announce his support for New Mexico Senator Hermann Schmitt, the former astronaut, in Schmitt's quiet campaign for a commission to set priorities for science.

Schmitt's commission would have certain advantages. For example, it could determine the correct balance between spending of the people's money for hybrid molecules to explore the causes of cancer and spending toward prevention of cancer, which is known to derive primarily from environmental pollutants. Again, a balance could be struck between spending to extract more nitrogen from the air and spending to develop a simpler method of obtaining nitrogen for crops—that is, by recovering the nitrogen from human and other animal wastes that now pile up in rural feedlots and municipal sewage treatment system dumps at a rate far exceeding anyone's present ability to dispose of them without polluting the streams.

Until the people are heard, there will continue to hang uncertainly before us, on hardly more than a thread of discipline, an incipient revolution in life—and not only human life. Just as we are beginning to talk about the rights of animals, we are faced with the necessity of thinking about the rights of microorganisms. It could hardly be otherwise. We are all relatives.

INDEX